Einstein Continued…

Einstein Continued…

The Missing Model of Motion

Martin O. Cook and David D. Miller

iUniverse, Inc.

New York   Bloomington

iUniverse books may be ordered through booksellers or by contacting:

iUniverse
1663 Liberty Drive
Bloomington, IN 47403
www.iuniverse.com
1-800-Authors (1-800-288-4677)

Because of the dynamic nature of the Internet, any Web addresses or links contained in this book may have changed since publication and may no longer be valid. The views expressed in this work are solely those of the author and do not necessarily reflect the views of the publisher, and the publisher hereby disclaims any responsibility for them.

ISBN: 978-1-4401-7621-0 (sc)
ISBN: 978-1-4401-7624-1 (hc)
ISBN: 978-1-4401-7623-4 (ebook)

Library of Congress Control Number: 2009936451

Printed in the United States of America

iUniverse rev. date: 10/02/2009

I want to know God's thoughts...the rest are details.

Albert Einstein

Contents

Preface

Timelessness provided the stimulus to think outside the box of *space-time* physics to explain the greatest mystery in science: quantum gravity.

Introduction

David and I began our philosophical and scientific studies together in 1998. In developing our theory of *omnipresence*—the *timelessness* of mass-energy through which consciousness flows—we concluded that even though the concept of *time* provided consciousness with the ability to sequentially order events, it didn't function as a physical reality weaved into the fabric of space. We specifically questioned the premise of *time dilation* in Einstein's Special Theory of Relativity. As one thought led to another, this inquiry eventually led to a quantum model for momentum, movement, relativity, and gravity, uniting the physics of Isaac Newton, Albert Einstein, and Werner Heisenberg. What makes this work significant is our absolute confidence in *timelessness*. Taking *time* out of Einstein's theories unravels the mystery preventing Einstein's work from being explained from a quantum perspective.

martin o. cook

Acknowledgements

Realizing the greatness of the people with whom I have been fortunate to forge friendships in the research and publishing of this book, I would like to acknowledge their involvement in the discussion of ideas contained visibly or invisibly in this book. A special thanks goes to David D. Miller for the numerous conversations that helped lead to the foundational ideas that the Missing Model of Motion is built upon. I also acknowledge his brilliant work in writing most of chapter 9, *The Illusion of Time*, for which I am extremely grateful. Along the way, David and I have chatted with other professionals and hobbyists in philosophy and physics. In these philosophical debates and scientific conversations, questions were asked, answers generated, and insights forthcoming that gave us the chance to dig in further and garner the motivation to publish this work. I also thank iUniverse for publishing our dreams into a reality for ourselves, our readers, and the scientific community. Lastly, I would like to acknowledge my wife, Wendy, for not only putting up with endless hours of conversation concerning the ideas in this book, but also for her insightful suggestions as editor of this book.

martin o. cook

PART I
Timelessness and Physics

Chapter 1

Relativity Continued...

○ ○

A complete, consistent unified theory is only the first step: our goal is a complete understanding of events around us, and of our own existence.

Stephen Hawking

After General Relativity, Einstein sought to unite his macro theory of gravity to the micro wonders of quantum mechanics. It makes sense that if the macro is made from the micro, there must be a valid connection between the two. Unfortunately, he died before he could accomplish this realization. The task we put before ourselves is to bridge the gap between these two divided theories, General Relativity and Quantum Mechanics. For this reason, our efforts are a continuation of Einstein's work.

Four of Einstein's important big breakthroughs, including three in 1905, began with imaginative insights that led to the solutions: stirring sugar cubes in coffee, conceptualizing light frequencies as quantum particles, visualizing time as a variable, and connecting gravity and acceleration. This last one being his happiest thought. Through these insights came the mathematical validations.

Mathematics is the language of science. It is more precise than words. Yet, with Einstein, mathematics was secondary, at least for the initial breakthroughs mentioned above. Philosophical insights developed into working mathematical theories. In the same vein, we attempt to continue Einstein's work into the quantum level, providing the foundational philosophical insights into quantum momentum, quantum movement, quantum relativity, and quantum gravity. Our insights are futile if they do not yield mathematical proof.

The Role of Philosophers

It may seem strange that a scientific theory of great significance is being presented in this book by people who were not trained at the university in the art of scientific and philosophical thought. In essence, we are adding insight into a 100 plus year old theory that withstood the onslaught of the Nazi regime and its attempt to discredit it. As hinted by Einstein—when one hundred or so scientists challenged in a formal statement one of his theories—if he were wrong, it would only take one.

As we question the validity of Einstein's use of *time* in the Special Theory of Relativity and the creation of *space-time* in the General Theory of Relativity, we admire Einstein's imaginative thinking and brilliant insights. In our opinion, he was a great philosopher and thinker who practiced mathematics and science.

The role of philosophers and thinkers is to question, to explore, to propose, to stretch the imagination beyond its present bounds, to turn things upside down, inside out, and right side up until the truth is made manifest for all to question and experience. Since our thinking is open to the potentials and consequences of *timelessness*, we are worthy candidates for this present task. The following is our current work on *timelessness and physics*.

Chapter Heading Quote: (Hawking, *A Brief History of Time*, 186.)

Chapter 2

A Fifth Grade Science Fair Experiment

○ ○

How can the same falling object travel two different differences?

The catapult of inspiration that led to the development of the Missing Model of Motion came from an unlikely source, helping my daughter with her fifth grade science fair project. Before I get into the project, I want to walk you through the events that eventually led to the Missing Model of Motion. It started back in 1998 when I met David at a part time job at Sears. Due to the nature of the job, answering incoming repair service calls, we had a lot of time to talk between calls on the graveyard shift. Within weeks before meeting David, I had come up with the idea of *omnipresence* from a spiritual and philosophical perspective, the emphasis for the follow-up book to this one. As David and I discussed and further developed the ideas of *omnipresence*, we were naturally led into the idea of *timelessness*. This led to a discussion that if *timelessness* was a reality, then *time*—as we experience it—was a concept of consciousness and not a physical actuality. This realization naturally led us to the work of Albert Einstein.

Special Relativity has changed the way people think about *time*. Time became the fourth dimension, being weaved into the fabric of space. The basic principle of Special Relativity is not that complicated. Special Relativity is the marriage of two contradicting principles, Galilean Relativity and Maxwell's constant for the speed of electromagnetic waves. It is the mathematics and consequences of Special Relativity that indulges in complexity. One important consequence of Special Relativity is *time dilation*. It stands in direct opposition to *timelessness*. If time dilation is a physical reality, then time exists and *timelessness* need not be further explored. On the other hand, if *timelessness* describes the perpetual state of mass-energy, then there is a problem with time dilation and Special Relativity because they depend on the physicality of time.

In trying to solve the time dilation dilemma, David and I worked out the basis for a theory that we called *The Propagation of Light independent of Uniform Motion*. This theory predicts that the *speed* and *direction* of light travels independent of the motion of the object emitting it. At first, we believed this in and of itself would answer the dilemma of the Michelson and Morley experiment. (Some encyclopedia examples of time dilation show the *speed* of light independent of the motion of the object emitting it, but the *direction* of the light—as demonstrated by a light pulse in a light mirror following the same pattern of a bouncing ball in Galilean relativity—remains dependent on the motion of the object emitting it—see Illustration—8 and Illustration—9.) I wrote a paper and sent it off to Galilean Electrodynamics, an organization I found on the Internet. To their kindness, they reviewed the paper and sent it back saying that our theory doesn't work. It didn't solve the Michelson and Morley dilemma. This left us perplexed. We were looking for a quick answer to our time dilation dilemma. We then postulated that the *speed* and *direction* of light traveling independent of uniform motion must impact the shape and movement of matter at the quantum level, meaning that the independent nature of light was the cause of the contraction measured in the Michelson and Morley experiment. We just couldn't explain how.

Insight From a Fifth Grade Science Fair Project

The year prior to the science fair I entered with Sarah, the school where I teach sixth grade conducted a science fair and picked out sixteen students to send to the next level, the district level. I participated as a first time judge at the district science fair. As judges, we were informed not to worry about noticeable parental help as long as the students were highly involved. In the end, my school only sent one student to the state level, a student who apparently had parental help. I left discouraged, not wanting to participate in a science fair again, not as a judge, teacher, or parent. This attitude permeated my thoughts going into the following year's science fair.

The following year, for the best interest of my students, I decided my school would participate in the science fair process again. My goal as a teacher was to learn how to help students advance to the regional level. My plan consisted of helping my daughter Sarah, who was in an accelerated 5th grade program at a different school, do an extraordinary project. We were going to do it on the impossibility of time travel but struggled to find an applicable experiment as a lead for the topic. Then the idea came to experiment with Galilean relativity. We would compare the distances of travel of the same falling object from two different inertial frames.

Because of the complexity of the question for her project, *how can the same falling object travel two different distances,* my struggle to find an answer helped me formulate quantum relativity. This eventually led to developing quantum momentum and quantum movement, and then quantum gravity. These four ideas fit together so well that they became four aspects to the same model: the Missing Model of Motion. When I began this project with my daughter, I didn't think it would lead to these revolutionary ideas that extends Einstein's work to the quantum level.

The funny thing about Sarah's project is that the judges from her school didn't forward it to the district level. I was astounded as only a dad and science teacher could be. Now what? Since my school was granted seventeen slots and I only planned on using sixteen, I received permission from the person over the science fair on the district level to send her project to the district level as a representative from my school. At districts, she won and earned the privilege of going on to the state level, which is held at Brigham Young University every year. Unfortunately, to our disappointment, she didn't place at the state level. Instead of focusing on Galilean relativity, I went for it all and wanted to show the obvious flaw in Special Relativity. Sarah wasn't quite ready to expound on ideas I was still in the process of formulating. My heart goes out to Sarah as she gave her best effort to please her dad. She learned as much as she could about ideas swimming around in my head.

Here is the question that we came up with for Sarah's science fair project:

Question

In Galilean Relativity, how can the same falling object travel two different distances?

Imagine two observers. Observer A is on a train going 100 miles per hour. Observer B is watching from a stationary position on the land. Imagine observer A drops a ball to the floor of the train. How will each observer see the path of the falling ball?

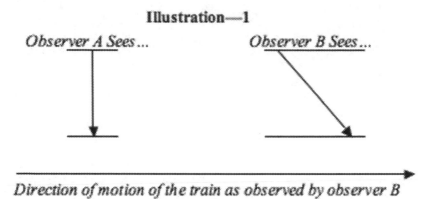

Illustration—1

Observer A Sees...   *Observer B Sees...*

Direction of motion of the train as observed by observer B

Illustration—2

Distance of A

Distance of B

Each observer measures the beanbag literally traveling a different distance.

The results are bizarre. The same falling ball actually travels two different distances.

We set up our experiment. Thomas, Sarah's older brother, would be in a car traveling at the uniform speeds of 0, 5, 10, 15, 20, and 25 miles per hour. At each speed, he would drop a beanbag at a designated spot 48 inches above the ground. He dropped it out the car window to the ground. Sarah would then measure the distance the beanbag fell from the designated drop off spot to the spot where it landed. When we performed the experiment, the results were interesting.

Illustration—3

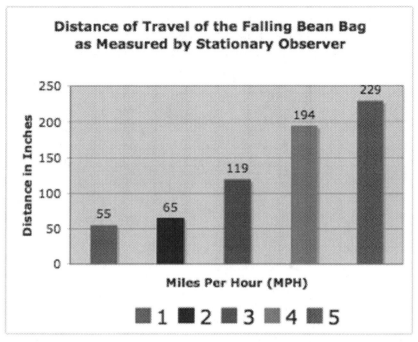

1 = 5 MPH 2 = 10 MPH 3 = 15 MPH 4 = 20 MPH 5 = 25 MPH

Even though the uniform speeds and distances measured were not precise scientific measurements, the results clearly demonstrated a pattern. With each increasing speed, the path of the beanbag increased for Observer B, Sarah. Yet, for observer A, Thomas, the beanbag fell the same distance of 48 inches every time on the assumption that he observed it falling in a closed environment. Because Thomas was dropping it from the window of a car, we could see the beanbag fall the two paths – the path observer B, Sarah, saw and the path observer A, Thomas, saw if he were in a closed environment. In the end, it was obvious to all of us who participated that the beanbag fell one distinct path, but depending on the relative position of the observers, the same falling beanbag could accurately be measured to fall different distances. How could the same falling beanbag be in two places at the same time, traveling through two different measurable amounts of space?

Drop a beanbag to the floor. As you watch it fall, ask yourself if it is possible for this beanbag to take two separate paths? Common sense dictates that the beanbag only travels one path. Yet, how is it possible that two observers in different inertial frames can measure different distances of movement or travel for the same falling beanbag?

As mentioned above, we concluded that the falling beanbag traveled one path. We then concluded that the differences in the observed distances were the result of the differing energy levels of the inertial frames. Observer A, Thomas, observed the falling beanbag from a higher energy level, the same energy level of the falling beanbag. Observer B, Sarah, observed the falling beanbag from a lower energy level of the falling beanbag. The greater the differences in energy levels, the greater the differences in measured distances of travel. A friend of mine pointed out that the distance Sarah observed was the simple formula of *A squared + B squared = C squared*. *A squared* is the distance Thomas saw the beanbag fall. *B squared* is the perpendicular distance from the dropping point to the landing point. *C squared* is the actual distance Sarah saw the bean bag travel from the time it was dropped to the time it hit the ground. The Pythagorean theorem:

Illustration—4

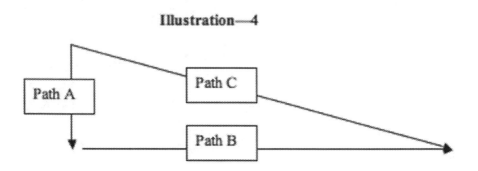

To us, this experiment proved that the overall energy level of the beanbag increased with each increased incremental change of the uniform motion speed. Another friend brought up an interesting point. For Thomas, if he were to drop the bag and then quickly reach down and catch the bag, he would only feel the effects of gravity on the beanbag. He wouldn't experience the increased energy level due to the increased uniform motion. Yet, if Sarah, from her lower energy level, reached out and grabbed the beanbag while it was falling, she would not only feel the influence of gravity, she would also experience the increased energy level of the beanbag caused by the increased energy acquired by the increased speed of the inertial frame from which it was dropped.

Here is another way of explaining what my friend was saying. If Thomas were on a train at the increased uniform motion and dropped the beanbag into a box of sand that was on the floor of the train, the indent created would be the effects of gravity only. This would be the same indent for all the differing inertial frames dropped in the same manner. Yet, if he dropped the beanbag from the train's increased inertial frame into a box of sand outside the train in Sarah's inertial frame, the indent would be larger because the beanbag would be carrying the effects of gravity plus the effects of increased energy of the inertial frame from which it was dropped. With each incremental increase of speed of Thomas' inertial frame, the indent would incrementally increase in Sarah's inertial frame.

The transfer of energy from a falling beanbag into a sandbox in the same inertial frame is the effect of gravity only because the sand and the beanbag are at the same energy level. When the beanbag is dropped into a sandbox at a lower energy level than the beanbag, the transfer of energy is greater because the sand in the lower energy level is absorbing the effects of gravity and the effects of the increased energy within the falling beanbag due to the increased energy level of the inertial frame from which it is dropped. It must absorb the gravity plus the extra energy of the increased motion. The insight was that the increased energy manifested by the beanbag was not a force acting on the beanbag, but rather, it was actual energy absorbed within the beanbag. With each incremental increase in the uniform motion, the beanbag literally absorbed more energy into its atomic structures. The measured differences of distances observed by Sarah were due to changes taking place within the beanbag on the quantum level.

This provided a major insight into quantum relativity of different inertial frames. The same object in different inertial frames contains differing amounts of energy, which translates into differing amounts of mass. Energy is absorbed or released with increases and decreases of movement. A simple science fair experiment led to new insights into Newtonian motion, Galilean relativity, Einstein's Special Relativity and eventually General Relativity. Before we explain these insights in greater detail, we will discuss the importance of the Michelson and Morley experiment to Special Relativity.

Chapter 3

The Michelson and Morley Experiment

○ ○

The Michelson and Morley experiment provided the basis for the Lorentz Transformation.

In the late 1800's, scientists believed *ether* was a motionless substance through which the earth moved.

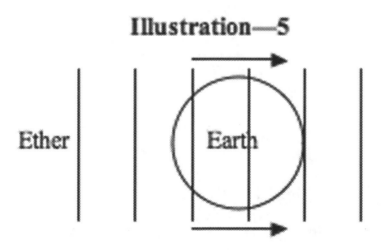

The Michelson and Morley experiment was specifically designed to measure the movement of the earth through this substance that many scientists believed permeated space. Just as energy travels through water in waves, scientists believed there needed to be a substance that permeated space that light waves propagated through. When Michelson and Morley failed to detect an ether wind as the earth moved through space, they originally believed their experiment failed. It was Albert Einstein that presented a theory eighteen years later that eliminated the need for ether to explain how light travels through space.

To say that the Michelson and Morley experiment wasn't necessary for Einstein's work on the Special Theory of Relativity would be like saying that Bernoulli's Principle wasn't important for the development of flight. The foundational equation for Einstein's theory comes from the Lorentz Transformation, an equation created as an attempt to account for the physical results of the Michelson and Morley experiment. Einstein uses *time* in conjunction with the Lorentz transformation to create Special Relativity.

Michelson and Morley believed the movement of the earth through this ether could be measured using the right instruments. The idea for the experiment was that the movement of the earth through the ether should easily be detected by measuring the interaction of light in the Michelson and Morley apparatus. As light entered into the apparatus, it would be separated, using a mirror and a clear window right next to each other. Then the divided light would travel at a right angle to each other, bounce off mirrors, and be brought back together. At this point, because the light should have traveled different distances, the observer could look through an eyepiece and check for interference patterns. This means that the wavelengths of the light that was divided and brought back together should no longer be lined up with each other but slightly shifted. The degree of the interference would indicate the speed of the moving earth through the stationary ether. We will recreate this with a diagram.

Illustration—6

The Michelson and Morley Apparatus

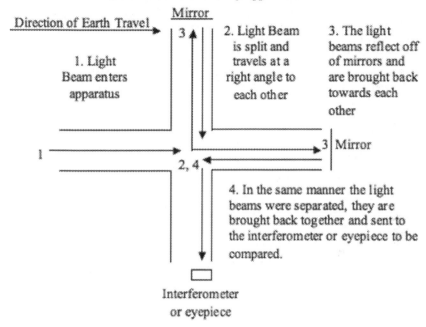

Direction of Earth Travel ➤

Mirror

3 ↑

1. Light Beam enters apparatus

2. Light Beam is split and travels at a right angle to each other

3. The light beams reflect off of mirrors and are brought back towards each other

1 ——————➤

2, 4

3 | Mirror

4. In the same manner the light beams were separated, they are brought back together and sent to the interferometer or eyepiece to be compared.

Interferometer or eyepiece

We hope the diagram is helpful in understanding the basic structure and purpose of the Michelson and Morley apparatus. After the light entered into the apparatus and the mirrors were adjusted, they turned the apparatus slowly and measured the results through the interferometer. The key to this experiment is the fact that the apparatus is resting on the earth, which is traveling 17 miles per second through space. Then the apparatus is slowly turned, changing the angles the light moves through the ether. The following will explain the premise for the experiment comparing two swimmers swimming at a right angle to each other in a current of water.

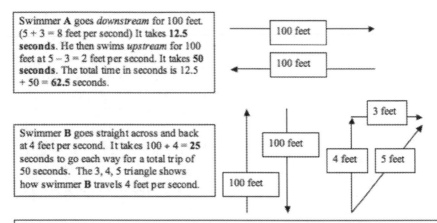

Illustration—7

Explanation for the Michelson and Morley Experiment

River flow – 3 feet per second

a. Each swimmer swims 5 feet per second
b. Each swimmer swims 200 feet
c. Swimmer A swims parallel to the bank 100 feet downstream then 100 feet up stream
d. Swimmer B swims straight across stream and straight back which is 100 feet each way
e. Swimmer A takes 62.5 seconds
f. Swimmer B takes 50 seconds

Swimmer A goes *downstream* for 100 feet. (5 + 3 = 8 feet per second) It takes **12.5 seconds**. He then swims *upstream* for 100 feet at 5 – 3 = 2 feet per second. It takes **50 seconds**. The total time in seconds is 12.5 + 50 = **62.5** seconds.

Swimmer B goes straight across and back at 4 feet per second. It takes 100 ÷ 4 = **25** seconds to go each way for a total trip of 50 seconds. The 3, 4, 5 triangle shows how swimmer B travels 4 feet per second.

The split light beam in the Michelson and Morley experiment should have acted the same as the two swimmers. When the beams were brought back together, one beam should have traveled farther then the other beam. There should have been interference patterns indicating that the two beams traveled different distances. Instead, there was a **null** effect indicating that the two beams traveled the same distance, even with the apparatus turning. How could this be?

No ether wind detected! Why?

Michelson and Morley thought their experiment was a failure because it failed to detect an ether wind. Even in this failure, they didn't question the existence of the ether. When they redid their experiment and obtained the same outcome, scientists attempted to explain the validated results.

George Francis Fitzgerald thought the apparatus could have contracted as it moved through the ether by external pressure applied on the apparatus. He developed an equation to explain the degree of the contraction needed to account for the measured results.

Hendrik Lorentz considered electrical and magnetic fields within the apparatus to account for the contraction—an internal cause. This didn't prove to be a viable explanation since it predated quantum mechanics. He, too, constructed an equation to account for the measured results. It is called the Lorentz transformation.

Neither explanation offered a workable scientific interpretation to explain the contraction needed to account for the physical results of the Michelson and Morley experiment. Yet, both scientists came up with a mathematical formula based on the results of the experiment that calculated the necessary contraction of the apparatus that would occur at different velocities to explain the null results of the experiment.

Using the formula created by Lorentz, *Albert Einstein* hypothesized that it was *time* that contracted. *Time* slowed down for faster moving objects. The Special Theory of Relativity was born.

In the end, something was happening that neither Michelson nor Morley expected and scientists such as Fitzgerald and Lorentz couldn't fully explain. Einstein came up with a solution that put an end for the need for an ether theory. He theorized that light travels in packets of energy he called quanta, disposing of a need for a medium through which light propagated. Then, using Lorentz's equation, he introduced the idea of space-time. In space-time, the speed of light was the same for all moving objects, maintaining James Clerk Maxwell's constant velocity for the speed of light in all inertial frames. In Special Relativity, all objects experience a time dilation proportionate to their movement through space. Time slows down as objects speed up.

Einstein's use of time to solve the Michelson and Morley dilemma contradicted our theory of *timelessness*. If space-time is a physical reality initiated by moving bodies through space, then *timelessness* is irrelevant. On the other hand, if *timelessness* explains the perpetual duration of mass-energy, then space-time provided a temporary solution to a quantum explanation. Einstein's use of *time* created the very stumbling block that prevented his discovery of a unified theory.

We will show how space-time isn't needed to explain quantum momentum, quantum movement, quantum relativity, and quantum gravity—the four legs to the Missing Model of Motion. The use of *time* to explain the unexplainable—Michelson and Morley dilemma—was a brilliant but temporary move.

Chapter 4

The Missing Model of Motion

○ ○

The missing model of motion answers the question that should have been asked centuries ago: How does mass move through space?

Newton's first law of motion is often stated that an object at rest tends to stay at rest and an object in motion tends to stay in motion with the same speed and in the same direction unless acted upon by an unbalanced force. From Newton to Einstein and beyond Einstein, physicists have taken for granted that objects move. They have attempted to explain the relative relationship between moving bodies with laws and theories such as Newton's Laws of Motion, Galilean Relativity, Special Relativity, and General Relativity. The one question not asked is why or how do objects move through space in the first place. The answer to this question is found on the quantum level of physics and is the missing link between macrophysics and microphysics. This missing link is the Rosetta stone for understanding quantum momentum, quantum movement, quantum relativity, and quantum gravity. It is the key to a unified theory that Einstein so desperately searched for in the declining years of his life.

During Newton's time, it would have been virtually impossible for him to develop a model for motion from a quantum perspective since quantum mechanics wasn't even in its prenatal stage. Einstein lived under similar restrictions since quantum physics didn't develop into a viable theory until the 1920's, after Special and General Relativities were developed.

Over the years, physicists have developed several perspectives for movement. We have Newtonian laws that are said to work well for objects moving at slow speeds. We have Einstein's Special and General Relativities which deal with faster moving objects, up to the speed of light, and large moving bodies such as planets and suns. We have Maxwell's constant for free flowing energies, and last of all, we have quantum physics, which explains movements within matter. Yet, none of these explain how *mass* moves through space in the first place.

The missing model of motion unites these somewhat opposing or separate views of movement into a single, viable model that explains the movement of *mass* through space from a quantum perspective, uniting macrophysics and microphysics, or in other words, General Relativity and quantum mechanics. The movements of large massive objects such as planets, solar systems, and galaxies are initiated and sustained on a quantum level.

The validation for the Missing Model of Motion came as individual parts unified into a single model. It began with an insight into quantum relativity. That insight naturally led into an insight into quantum momentum and quantum movement. With these three pieces of the puzzle in place, I knew that quantum gravity was only a few connective thoughts away. It was the most difficult of the four pieces to formulate but when the ideas started to flow, quantum gravity fit nicely into the already formulated pieces of quantum relativity, quantum momentum and quantum movement. What helped make the push from the first three pieces to this last piece is that I had read a year or two previously that quantum relativity has yet to be discovered and would probably be developed within the author's lifetime and would be the link or catalyst to discovering quantum gravity. This provided the initiative to keep pushing towards an understanding of quantum gravity. Because these four pieces fit together so nicely, I am confident in presenting the Missing Model of Motion.

Chapter 5

Quantum Momentum

○ ○

Quantum Momentum explains the perpetual momentum of mass through space.

What was the simple error that Einstein made that prevented him from discovering a quantum theory for relativity and eventually a quantum theory for gravity? Einstein didn't define quantum momentum. He didn't ask the monumental question: How does mass move through space? In 1905, what was known about quantum mechanics? As a theory, it was in the prenatal stage. Einstein was stuck in a Galilean mindset as he composed Special and General Relativities. He didn't question or answer *how* macro objects move through space. Newton's first law of motion states that a body in motion tends to stay in motion unless an unbalanced force acts upon it. Why does mass tend to stay in motion unless a force acts upon it?

I am going to take a few steps backwards to the time of Isaac Newton, before he was bladed a Sir. His first law of motion from a quantum perspective yields quantum momentum and leads to an understanding of quantum movement, quantum relativity, and quantum gravity. The following should help in conceptualizing quantum momentum.

All Mass is in Motion through Space

Mass is always in motion through space. It is never at rest. Yet, from a macrophysics perspective, we see things that appear to be at rest on the surface of the earth. If we were to see the same things from a quantum perspective, we would see electrons orbiting around nuclei billions of times in a millionth

21

of a second. Even objects that appear to be at rest are comprised of moving parts.

When resting on the earth, mass seems to lack motion or movement. Yet, the very computer I am using to type this book is moving through space at seventeen miles per second as it rests upon the earth that is moving at this speed. Even objects that appear to be at rest are moving through space.

The source and cause for the movement of mass isn't starting from a non-moving position and then speeding it up, but on the contrary, it is energies moving at the speed of light, combining to create forces. These forces eventually combine into atoms and compounds. As energies combine to form mass, the combined mass as a whole moves through space at speeds less than the speed of light, but the energies that comprise this mass continue to travel at the speed of light. Thus, the slower speed of mass as a whole is comprised of confined energies moving at the speed of light. I use *energies* to imply the different frequencies of energies that combine into forces that create the phenomenon of *mass-energies in motion* or just *mass in motion*. From this point forward, I will refer to *mass in motion* as either *confined energies* or just *mass*. *Mass* as *confined energies* gives meaning to Einstein's equation $E=MC^2$. *Mass* is nothing more than the energies that comprise it. The equivalency between freely moving energies and *mass* as *confined energies* creates the framework for quantum momentum.

Confined energies are the source and cause for the *motion of mass* through space. *Mass* is *confined energies* moving through space. If you remove the energies from *mass*, there is nothing left over—no spare parts. The *confined energies* are the *mass*. Thus, *confined energies* are the momentum of *mass* through space.

It is important to visualize that the motion of *mass* through space is the result of the movement of the confined energies. The energies are already traveling at the speed of light through space. As you combine them into synchronized orbital patterns, the energies are still moving at the same speed, but the overall speed of the synchronized orbital patterns as a whole moves through space at slower speeds than the energies comprising the orbital patterns. This is *mass*. Since the energies are already moving through space, the synchronized orbital patterns of these energies are the cause and source of the momentum of *mass* through space. In other words, *mass* is the synchronized orbital pattern of energies, which is the source and cause of its movement through space.

At first, the idea that the momentum of *mass* through space is the synchronized movement of the *confined energies* from which *mass* is comprised may be difficult to visualize, but as the pieces to the Missing Model of Motion come together, the process of visualizing *mass* from this perspective becomes very natural. It becomes difficult not to visualize the momentum of *mass* through space from this perspective.

What makes this concept difficult to grasp is we continually see *mass* at rest on the earth. The computer I am typing on does not appear to have any momentum. This is because the computer is in a continuous state of acceleration towards the earth—gravity. As it sits on my desk, it is temporarily experiencing terminal velocity. Its momentum is the same as the momentum of the earth through space. If the effects of gravity could temporarily be shut off, the momentum of the computer through space would be more apparent. It would continue to move through space at the same momentum of earth without being connected to the earth because it would no longer be accelerating into it. If you gave it a push, it would slowly drift away from the earth, accentuating its momentum through space.

A good way to visualize quantum momentum is to think about objects in momentum in the space shuttle while it is orbiting the earth. When the space shuttle is in a freefall, the momentum of objects can clearly be observed. When an astronaut pushes an object, such as a bag of water, the bag of water continues in the same direction until it bumps into a wall of the space shuttle. Objects that the astronauts place in the air next to them—such as a toothbrush—stay in the same place. This is because it has the same momentum in space as the astronaut. Their momentums are synchronized, moving at the same *speed* and in the same *direction*.

Momentum levels of *mass* do not change—a brilliant call by Newton—unless the *confined energies*—the source and cause of its motion—are acted upon, (externally or internally). Changes to the *confined energies*, externally such as a push or internally such as the absorption or emission of *energy*, causes movement—changes to momentum. From this point forward, I will refer to changes in momentum as *movement*.

This is the first principle of the Missing Model of Motion: The momentum of *mass* through space is the synchronized movement of the *confined energies* from which it is comprised. Thus, the speed and direction of *mass* through space is constant just as Newton observed unless unbalanced (external or internal) forces act upon it.

Chapter 6

Quantum Movement

○ ○

Changes in Momentum

Quantum momentum explains how *mass* remains in perpetual motion unless acted upon. Quantum movement explains momentum changes from a quantum perspective and deals specifically with the absorption and emissions of energies within *mass*. Because quantum movement requires an introduction and explanation of *energy levels* and *orbital patterns*, which topics are introduced in quantum relativity, quantum movement will be explained in chapter 7 on quantum relativity.

I inserted this very short chapter to stress the logical sequence from quantum momentum, to quantum movement, to quantum relativity, and finally to quantum gravity. Even though these are four interconnected parts of the same model—the Missing Model of Motion—it is important to see the role of each part. Quantum relativity is a combination of quantum momentum and quantum movement.

Chapter 7

Quantum Relativity

○ ○

What will be the impact of quantum relativity on Special Relativity?
Time *will tell.*

What is quantum relativity? It is an explanation of Galilean relativity from a quantum perspective. Quantum relativity explains Special Relativity without the use of *time* by introducing energy levels and orbital patterns on an atomic level. It is the third of four legs that explain the Missing Model of Motion.

Quantum Relativity

In Galilean relativity, Galileo learned that if a person were in a closed cabin of a ship moving at a uniform speed, he wouldn't be able to tell his relative motion to land by an appeal to the laws of physics. If he dropped an object, it would fall straight down. If he threw it up in the air, it would go straight up and straight down. In all, there was nothing he could do to determine his relative speed to land by applying what he knew about the laws of physics. This observation of his is on a macrophysics level, physics we can see with our eyes. Even though he observed no differences in the laws of physics on a macrophysics level, would the same hold true on the quantum level? Are there measurable differences on the quantum level for different inertial frames?

Einstein spent much of his adult life searching for a unified theory that would resolve the contradiction between General Relativity with quantum mechanics. Try as he might, he could not accomplish this task. Einstein's assumption that the laws of physics are the same for all inertial frames created the very roadblock or barricade that prevented him from achieving his goal of a unified theory. From a macrophysics perspective, all the laws of physics appear to be the same for all inertial frames, but from a microphysics or quantum perspective, there are differences.

The science fair project that Sarah, Thomas, and I did provided the spark of insight needed to understand quantum relativity. The question: *How can the same falling object travel two different distances?* The answer to the question pointed to energy levels. Energy levels will be defined later. For now, energy levels refer to the orbital patterns that are the cause and source of the momentum of *mass* through space, determining the speed *mass* moves through space.

With each incremental increase of speed to the beanbag as it traveled in the moving car—*relative* to earth's inertial frame—the more overall energy the beanbag acquired and contained. For example, if you dropped a beanbag into a pile of sand outside of a *non-moving* car, the beanbag would displace a certain amount of sand. The energy acquired by the falling beanbag due to the effects of gravity is transferred to the sand at impact. Now, if the car was moving and you dropped the beanbag so it hits the same sandbox that is outside of the car, falling at an angle to the ground due to the movement of the car, the overall amount of displacement of sand caused by the falling beanbag would be increased. The beanbag literally contained more energy to be transferred to the sand on impact, creating a greater displacement of sand.

I postulated that this increase of energy was an actual increase of energy that was added within the beanbag. Thus, any change in *momentum* is a change in the overall energy level, the overall amount of energy comprising that *mass*. This explained how two different observers could measure the same falling object to travel two different distances. The difference in distances resulted from the differences in the energy level of each observer and the energy level of the beanbag.

From a macrophysics perspective, the differences in distances is shown by the formula $d=r \cdot t$: *distance = rate • time*. The time that it took for the beanbag to fall to the ground was the same for both observers due to the same gravitational effects on the beanbag. The rate at which it fell was different for each observer, yielding two different distances. For each observer, the beanbag was traveling at different speeds. How can one beanbag travel at two different speeds, yielding different distances, simultaneously?

From a quantum perspective, rate refers to the energy level of the orbital patterns comprising the *mass*. The energy level of the beanbag was the same as the energy level as Thomas and a different energy level than Sarah. The energy level of each observer in relation to the energy level of the beanbag determined the observed speed and distance of the falling beanbag. Even though, according to each observer's measurements, the beanbag traveled at two different speeds and went two different distances, the beanbag took only one absolute path through space. These differences in speed and distance are now accounted for from a quantum perspective by comparing the different energy levels from which each observer watched the beanbag fall in comparison to the energy level of the beanbag. Since Thomas and the beanbag were at the same energy level or inertial frame, the beanbag fell straight down as observed in Galilean relativity. Sarah, at a lower energy level or inertial frame, sees the beanbag falling from the higher energy level or inertial frame. Her measurements are the results of the differing energy levels between her and the beanbag.

Somehow, the faster the beanbag moved relative to the earth's inertial frame, the more energy it acquired. Thus, the energy level of the beanbag changed when the inertial frame that carried it changed. As the car accelerated to go from one inertial frame to the next, say from 5 mph to 10 mph, the beanbag literally acquired more energy within its atomic structures. The observer at a lower energy level than the beanbag experienced the momentum of the beanbag through space differently then the observer at the same energy level as the beanbag. The differing energy levels accounts for the differing distances and speeds. Without differing energy levels, *masses* would not experience differing momentum speeds through space.

In Galilean relativity, from a macrophysics perspective, there appears to be no differences in the laws of physics for the same *mass* in differing inertial frames. A ball always falls directly towards the earth's surface in all inertial frames. From a microphysics perspective, there are differences in the energy levels of the orbital patterns of the same *mass* in different inertial frames. The ball has differing amounts of energy inside of it in different inertial frames. Changes in momentum—*movement*—are changes in energy levels of the orbital patterns comprising the *mass*.

Quantum Movement

Orbital Patterns

What do I mean by energy levels and orbital patterns? Energy levels and orbital patterns are the driving force behind the plasticity of mass. They account for changes within *mass*. A change within *mass* is a change in momentum or *movement*. Energy levels and orbital patterns are the source and cause of the momentum of *mass* through space. Thus, energy levels and orbital patterns are the confined energies that comprise mass.

It is important to note that orbital patterns are a description of a process. It explains the synchronized movements of *confined energies* that account for the movement of *mass* through space. Whether in a particle, atom, molecule, compound, or mix, the *confined energies* share an *order of movement* that is the source and cause of the *momentum of mass* through space. It is easier to visualize momentum of *mass* in space rather than on earth. When I need to visualize quantum momentum, I usually visualize objects in a free falling space shuttle. Then I can visualize how *mass* is always in momentum through space such as an orange appearing to float from one astronaut to another. The reason it is difficult to visualize the perpetual momentum of *mass* on earth is because *mass* is either at rest on the earth's surface, sharing the same momentum of the earth, or it is accelerating towards earth's surface, the effects of quantum gravity or changes to momentum.

Orbital patterns are the synchronized *order of movement* of the *confined energies* that are the source and cause of the momentum of *mass* through space. Think of it this way. If an electromagnetic wave moves in a straight line from point A to point B, it would travel at a speed of 186,300 miles per second to cover the distance between the points. If that electromagnetic wave were to travel in a circular path as it moved forward from the same point A to the same point B, the distance of travel of the electromagnetic wave would be increased as well as the time it takes to go from point A to point B. Even though the speed of the electromagnetic wave would remain constant, the overall distance of travel and the time it takes to travel that distance would increase due to its movement in an orbital—circular—path. A change in the size of its orbital path as it travels through space would either increase or decrease the distance and time it would take for the electromagnetic wave to go from the same point A to the same point B, depending on whether this change was a larger or smaller orbital—circular—path. *A change in the size of the orbital path would change the speed of its momentum through space.* I am not insinuating in the least that an electromagnetic wave can travel in an orbital path freely through space. I am saying that *confined energies,* from which *mass* is comprised, travel in orbital patterns through space. This is the source and cause of the momentum of *mass* through space.

From this point forward, I will refer to orbital paths of confined energies as *orbital patterns.* I use the word *patterns* because the momentum of *mass* remains consistent unless it is acted upon by an unbalanced force. A pattern of movement is maintained that keeps the *mass* in momentum through space. *Movement* is initiated when orbital patterns are disrupted.

Energy Levels

The size of the orbital patterns is determined by its energy level. The unique quality about *mass* is its ability to absorb and emit energy. This quality is the driving force for *quantum movement* and *quantum gravity* as will be explained. The energy level of orbital patterns can increase by the absorption of energy or decrease by the emission of energy. The energy level and orbital patterns are inseparably connected. *A change in orbital patterns is a change in the energy level. And vice versa, a change in the energy level is a change in orbital patterns.*

Imagine *mass* as orbital patterns. Then imagine the orbital patterns in concert with each other as the source and cause of the momentum of *mass*—speed and direction—through space. A change in the energy level is a change in the orbital patterns, which is a change in the momentum or speed of that *mass* through space, (the basis for quantum gravity), and visa-versa, a change in momentum is a change in orbital patterns, which is a change in the energy level. Momentum is maintained by the energy level of the orbital patterns of *mass*. *A change in momentum is a change in the energy level and orbital patterns, and a change in the energy level and orbital patterns is a change in momentum.*

A change in the orbital patterns of *mass* is a change in the energy level of that mass, such as when two particles collide. At the moment of impact, orbital patterns are disrupted. This means that the frequency of the energy comprising the orbital patterns shift. A frequency cannot shift without absorbing or emitting energy to correlate with the shift. This is very important because it is the driving force for the elasticity of *mass*. As orbital patterns are altered, the accompanying energy level is altered by the absorption or emissions of *energy* to exactly coincide with the degree that the orbital patterns were altered. This changes the orbital patterns and their corresponding energy level, changing the momentum of that *mass* through space until there is another disruption to the orbital patterns or energy level of that *mass*. Changes in orbital patterns with its corresponding change in energy level by the absorptions or emissions of energies transpire at the speed of light. *The equilibrium of mass is the momentum of that mass through space.*

It is easy to surmise that an increase in momentum requires the absorption of energy to compensate for the accompanying changes in orbital patterns. And vice versa, a decrease in momentum requires the emissions of energy to compensate for the accompanying changes in orbital patterns. The faster *mass* moves through space, the more *mass—energy—*it acquires. Why is this?

An increase or decrease in the momentum of *mass* through space is determined by the absorption or emission of energy in relation to the orbital patterns. The number of atoms that comprise the *mass* does not changed; it is the amount of energy contained within the atoms that changes. This has to do with the frequency and wavelength of the energy that comprise the orbital patterns of the *mass*. Higher frequencies with shorter wavelengths have more energy than lower frequencies with longer wavelengths in the electromagnetic spectrum. Thus, more energy is required to decrease the synchronized size of orbital patterns in order to increase its momentum through space. (For a more detailed explanation, see Appendix 1 titled *Gravitational* Energy.)

Isn't it interesting that in Special Relativity, as mass increases in speed, its size contracts but its overall mass increases? Remember this thought as I further explain the energy level of orbital patterns.

In summary, the orbital patterns of the energies that make up *mass* are synchronized as the source and cause of the momentum of *mass* through space. Changes in orbital patterns with its corresponding energy level are the source and cause of *quantum movement,* a change in quantum momentum. Up to this point in physics, the source and cause of the momentum and *movement* of *mass* through space from a quantum perspective has been taken for granted.

The idea that *mass* is confined energies of synchronized movements, which is the source and cause of its momentum through space, may still be difficult to visualize, but as the pieces to the Missing Model of Motion are further linked together, the visualization as such will become very natural. It actually becomes difficult not to visualize the momentum and *movement* of *mass* through space from this perspective.

Quantum Relativity

How did the same falling beanbag travel two different distances? The answer is quite simple. It didn't. The differences in distance had little to do with the beanbag and had everything to do with the energy level—inertial frame—of each observer. Think of the following experiment done on the space shuttle freefalling around the earth. Person A is moving from one side of the space shuttle to the other side of the space shuttle. When he reaches person B, who is stationed in the middle of the space shuttle, he lets go of the beanbag. The beanbag will continue to travel with person A because their momentums are the same. They are at the same energy level, and their orbital patterns are synchronized even though they are not bonded together. On the other hand, person B watches as the beanbag slowly moves away at the same speed that person A is moving away. Even though the beanbag travels only one distinct path, each observer measures a different length of travel from his or her immediate perspective. Even though each observer can account for a

distinct difference in the distance of travel of the beanbag through space from their immediate perspective, the beanbag's actual movement through space was absolute.

Galilean relativity is nothing more than differences in the energy level of orbital patterns of the observers and the *mass* being observed. If Galileo could have measured the energy level of the orbital patterns of any object in his boat cabin, he could have compared them to previously measured energy level of orbital patterns of anything attached to land. The measured difference would account for the increased speed of his cabin in relation to the land.

All mass travels an absolute path through space.

Acceleration and deceleration of *mass* are temporary stages between the equilibrium momentum patterns of inertial frames. Accelerations and decelerations of *mass* are accompanied by absorptions and emissions of energies to accommodate for the changes in the synchronized orbital patterns and the energy level required to maintain them. This accounts for the perpetual momentum and movement of *mass* through space.

The Light/Time Dilemma

Now we advance the question of the same falling beanbag traveling two different distances to the propagation of emitted light. How can the same emitted light pulse travel two different distances? Steven Hawking put it like this:

> Since the speed of the light is just the distance it has traveled divided by the time it has taken, different observers would measure different speeds for the light. In relativity, on the other hand, all observers *must* agree on how fast light travels. They still, however, do not agree on the distance the light has traveled, so they must therefore now also disagree over the time it has taken. (The time taken is the distance the light has traveled – which the observers do not agree on – divided by the light's speed – which they do agree on.) In other words, the theory of relativity put an end to the idea of absolute time! (Hawking, 1996, 21-22.)

To answer the dilemma proposed by Stephen Hawking, we ask the following question: *What is the major difference between a light pulse and a beanbag?* The beanbag can be at rest in multiple inertial frames from a macrophysics perspective. On the other hand, an emitted light pulse is never at rest in any inertial frame. This means that an emitted light pulse operates differently then the laws of motion for *mass* as established by Newtonian physics and Galilean relativity. For example, it is common knowledge that the *speed of light* is unaffected by the *speed* of the mass emitting it, but the *speed of an object* is directly affected by the *speed of the mass* from which it is released. This major difference is due to the fact that in Newtonian physics, objects can acquire a state of rest within an inertial frame—same energy level of orbital patterns—whereas a freely moving light pulse never finds rest within any inertial frame.

This brings up the two postulates that Einstein uses to build his Special Theory for Relativity. His first postulate states that the laws of physics are the same for all inertial frames. This correlates with Galilean relativity wherein a person in a closed cabin of a ship cannot tell his relative motion to the land by appealing to the macro laws of physics. He follows that up by stating that the speed of light is the same for all moving objects. This correlates to Maxwell's mathematical discovery that electromagnetic waves travel at a fixed, unchanging velocity. When Einstein combined the two postulates, he theorized that the speed of light was constant—same—for all inertial frames. In other words, the speed of light was the same for all moving objects—a person walking, a flying airplane, and a speeding rocket. I will explain why this assumption is false and how the speed of light is constant for all moving objects without the need to manipulate time.

Let's quickly diagram the problem.

Illustration—8
Light Clocks

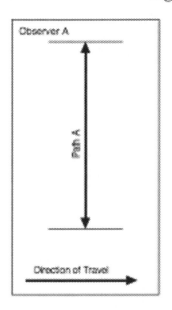 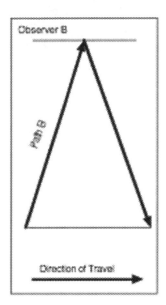

If *observer A* is traveling in a train with a hypothetical light clock, the light will move in an up and down motion in reference to *observer A* (Path A). Yet, *observer B*, who is stationary to the motion of the train that is carrying *observer A* and his light clock, will observe the path of the light moving at angles in the direction of the uniform motion of *observer A* (Path B). The distance of *path A* is different than the distance of *Path B*. Yet, from each person's perspective, they are both correct. This is quite a paradox if the speed of light is constant and yet travels two different distances. (In the case of the beanbag, the speed of the beanbag was different for each observer because each observer observed from differing energy levels. This accounted for the different distances that the beanbag traveled as measured by each observer even though the beanbag only traveled one distinct path through space.) Thus, Einstein's brilliant idea was to incorporate the concept of *time* to answer this paradox. *Observer A* and *Observer B* experience the speed of light at its constant speed, yet, they each experience *time* differently in order to compensate for the differences in distances traveled by the same light pulse. Illustration—8 is a classic example used to explain time dilation of Special Relativity in many up to date encyclopedias.

In illustration—8 if you replaced the light pulse with a bouncing ball, this would be an example of classic Galilean relativity for which we have already provided an explanation on a quantum level. However, the explanation to the light pulse dilemma is found in the differences between a freely moving light pulse and the inertial momentum of *mass*.

David and I spent many hours discussing this paradox. We came to the conclusion that a light pulse travels independently of the uniform motion of the *mass* from which it is emitted, realizing this consequential argument:

> *If the speed of light is independent of the motion of the mass emitting it, then its direction or path should also be independent of that motion.*

This answered the light pulse dilemma. The speed and **direction** of emitted light pulses are unaffected by the motion or momentum of the object emitting it. *Time* was not needed to explain the paradox. An object of *mass*, such as the beanbag, which can be at rest in an observed inertial frame, must be treated differently than freely moving light pulses, which are never at rest in any observed inertial frame. In other words, freely moving light pulses cannot be regarded in the same manner as *mass* in respect to inertial frames.

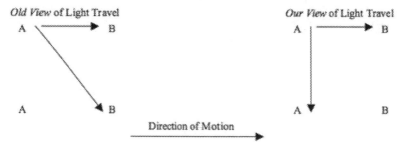

Illustration—9
Different Views for the Propagation of Light Emitted from Mass in Motion

A represents uniformly moving *mass* in the direction of *B; A* also represents the emission of a light pulse perpendicular to the motion of the *mass* emitting it. The *Old View* demonstrates light's direction or path of travel *dependant* on the motion of the *mass* emitting it—a continuation of Galilean relativity. The *Our View* demonstrates light's direction or path of travel *independent* of the motion of the mass emitting it.

Just as the *speed* of light is unaffected by the motion of the *mass* emitting it, the *path or direction* it travels after its emission is also unaffected by the motion of the *mass* emitting it. This changes everything postulated about freely moving light energies in relation to differing inertial frames of *mass*. The independence of the direction and speed of light from the inertial frames of *mass* is the foundation for the Missing Model of Motion. This *new view* of light is the mechanism that drives changes in quantum momentum, the source and cause of quantum movement, quantum relativity, and quantum gravity. Light's *speed* and *direction* are not perpetually affected by the motion—momentum—of the *mass* emitting it.

The momentum of *mass* from which an object is released perpetually affects that object's momentum—its speed and direction—like an object dropped from an airplane or a ball dropped in a uniformly moving train. The reason for an object being perpetually affected by the momentum of the *mass* from which it was released is because it shares the same energy level of orbital patterns as the *mass* from which it was released, just like the example of the astronaut in the space shuttle who lets go of a beanbag or toothbrush. In the case of an object moving away from the *mass* from which it was released—such as a suitcase falling from an airplane—the energy level of orbital patterns of the suitcase are also being accelerated towards the earth's surface by the absorptions of energies from that direction—quantum gravity.

I remember someone telling me how Special Relativity was demonstrated in a college class. One student took a piece of chalk and perpetually drew a line going up and down while staying in one place. Then a second person did a similar thing; they walked along the chalkboard while doing the up and down motion with the chalk. It supposedly demonstrated how light traveled different distances in different inertial frames even though they were going the same speed—same rate of up and down motions. Unfortunately, this isn't how light works. The speed of light is not the same for differing inertial frame. Light operates independently of all inertial frames.

Once light is emitted from the confined energies of inertial mass, its *speed* and *direction* of travel moves independently from the momentum of the mass emitting it. How science ever gave light relativistic qualities similar to a bouncing ball in Galilean relativity is beyond my ability to presently comprehend. This false assumption of the movement of energies within inertial frames prolonged the gap between macrophysics and microphysics. It is the independent nature of energies from the inertial movements of *mass* that drive the elasticity and changeability of *mass* on a quantum level, dictating the absorption and emissions of energies as momentum changes occur.

At this point, we thought the independent nature of light in and of itself answered the Michelson and Morley dilemma. I typed up a paper and sent it off as I had mentioned earlier. I received the response saying that this theory in and of itself did not account for the dilemma of the Michelson and Morley experiment. Feeling strongly that we were on the right track, we continued to brainstorm. I suggested that this somehow must be the cause for the contraction calculated by Fitzgerald and Lorentz. Somehow, energies contained or confined within the atomic level, whose speed and direction respond independently of the inertial frame of the mass containing them, accounted for the contraction of *mass* at differing speeds. At this point, I had no idea that the confined energies that constitute *mass* were the cause and source of its momentum through space. Then I did the science fair project with Sarah as discussed in detail in an earlier chapter, which brought renewed insight into this paradox.

As insights came, I realized that Einstein was right about one of his postulates: The speed of light is the same for all moving objects. There is consistency in the universe. Yet, my explanation as to how this is possible is different. Einstein viewed this idea from a macrophysics perspective such as a moving car or rocket, (mass as a whole and not from its parts), and that the speed of light would always pass a moving car or speeding rocket at its constant velocity of *c*, (*c* refers to the speed of light). He justified this by calculating that as a macro object was in motion, its *time* reference slowed down to the point that light passed it at its constant speed. Thus, every moving object contained its own *time* in reference to the speed of light so that Maxwell's equations were not nullified. In this way, the speed of light remained constant for all moving objects and in all inertial frames. This was a brilliant move. The fundamental equation upon which this is based came from the observable results of the Michelson and Morley experiment as calculated in the Lorentz transformation. Even though Einstein didn't base his theory as an attempt to answer the Michelson and Morley experiment, his use of the Lorentz transformation formula leaves no doubt that his theory is directly influenced by this experiment. No Michelson and Morley experiment means no Lorentz transformation formula. As we will discuss later, Einstein's use of the Lorentz transformation and his creative use of *time* bring validity to his descriptive equations.

The Missing Model of Motion addresses this very idea that the speed of light is the same for all moving objects. Instead of appealing to macrophysics and the observable motion of objects, I appeal to microphysics and the *confined energies* on the quantum level. Every atom is made up of energies, which travel at the speed of light in orbital patterns. This constitutes the movement of any *mass* through space. When light passes any macro object, it is traveling at the same speed of all the energies from which the macro object is constructed. Instead of comparing the speed of light to the macro object as a whole and then using *time* as the variable to keep the speed of light constant for all moving macro objects, all objects on a microphysics or quantum level are made up of orbital energies already traveling at the speed of light. The speed of light moving freely through space is the same as the speed of light—

energies—comprising mass. This is possible because there is nothing about *mass*—confined energies in motion—that isn't moving at the speed of light when separated into its simplest energies. Einstein's equation that says $E=mc^2$ literally means energies and *mass* are equivalent. *Mass* is nothing more than *confined energies* in motion. Not taking this literally has been the great divider between macrophysics and microphysics. The two are really one.

If the *speed* and *path* of energies are unaffected by the motion of the mass emitting it and if *mass* at the quantum level is comprised of energies moving at the speed of light, (even electrons which move at eighty-five percent of the speed of light are made up of photons which move at the speed of light), then energies that comprise any inertial frame will maintain constant velocity with itself, including any free flowing energies traveling in the universe. Instead of comparing a photon passing a moving rocket and using *time* to slow down the rocket to maintain light's constant velocity, I compare the photon passing the energies that comprise the *mass* of the rocket at a quantum level to maintain the constant velocity of light throughout the universe. It is comparing light to the energies that constitute *mass* rather then comparing it to our macro perspective of that *mass*.

Are there naturally existing particles independent of energies that travel less then the speed of light? On the microphysics or quantum level, energies in their purest form travel at the speed of light, whether free flowing or confined as *mass*. Electrons are known to travel at about eighty-five percent of the speed of light. Yet, the photons that make up an electron are traveling at the speed of light. It is because of these photons comprising the electron that the electron even moves at all. It is amazing to think that all objects, including everything we see on the macrophysics level, could be comprised of energies that are constantly traveling at the speed of light. Here, in our opinion, is the great miracle. Even though *mass* is made up of energies, the energy level of orbital patterns of those energies enables *mass* to obtain a state of rest within any inertial frame. That is, they share the same energy level of orbital patterns with other macro objects that comprise the same inertial frame. Thus, the energy level of orbital patterns is the key for understanding quantum relativity.

The Absorptions and Emissions of Energies

What is energy and how does it move through objects? As I try to explain energy to my sixth grade students by making waves with a rope, I ask them: *What is it that is traveling through the rope?* Someone will usually give the answer *energy*. But what is energy? How does it make the rope move? This question of how energy moves through the rope perplexed me. A reaction takes place, but why, or even more important, how?

An in-depth look into quantum relativity explains the answer to this question. The observed movement of the rope is caused by a chain reaction on the atomic level. Before the rope is whipped, the orbital patterns from which the rope is comprised are in par with the orbital patterns of an orbiting earth through space. As the rope is whipped, the orbital patterns comprising that part of the rope change. The orbital patterns react immediately to this disruption because energies do not skip or miss a frequency. This disruption in the orbital patterns from which the rope is comprised creates a chain reaction because any change in orbital patterns is a change in the energy level, a process of emitting and absorbing energies on the quantum level to maintain equilibrium. This chain reaction creates the resulting and observable movement of the rope. *In other words, emissions and absorptions of energies in mass account for changes in momentum in the rope, which is initiated by the disruption of the orbital patterns of the energies comprising the rope.* Because the atoms of the rope are bonded or held together, the disruption of one orbital pattern causes a disruption to another and another, creating a chain reaction throughout the entire length of the rope. What is happening at the speed of light on the quantum level, a massive exchange of energies due to orbital changes within each atom, we see as a comparatively very slow moving wave through the rope on a macrophysics level.

The driving force of this chain reaction is the independent nature of the energies comprising the orbital patterns. Since energies have a constant velocity, any disruption to *mass* is a disruption to the orbital patterns comprising that mass. This creates a shift in the frequency of the energies comprising the orbital patterns. The shift simultaneously initiates the emission or absorption of energies since each frequency is a finite amount of energy. Thus, a change in frequency is a change in the energy level, and visa versa, a change in the energy level is a change in frequency.

Thus, any change in momentum—a *movement*—is a change in orbital patterns. This creates a chain reaction in the absorptions or emissions of energies, whether that is an increase or a decrease in the overall energy required to maintain the new orbital patterns. The emissions and absorptions of energies on a quantum level, triggered by disrupted orbital patterns, causing changes in momentum, is the actual source and cause of the rope moving. Again, what we see as a relatively slow moving wave are the chain reaction of momentum changes caused by the absorptions and emissions of energies transpiring at the speed of light.

As two independently moving particles bump into each other, there is a change in their overall energy levels as orbital patterns adjust to that collision. The energy level of orbital patterns always maintains equilibrium even if it requires that bonds sever and *mass* break apart such as the case when a mirror falls to the floor and breaks into multiple pieces. When I first started writing this book, I thought that a change to the energy level was a change to the orbital patterns, and vice versa, as if they were two separate aspects influencing each other. At the time, it didn't occur to me how they were one in the same process. Then I realized that the energy level of orbital patterns coincide with electromagnetic frequencies. This means that the orbital patterns are comprised of electromagnetic waves. A change to an electromagnetic frequency is a change in the energy of that frequency and visa versa. A disruption of a frequency caused by movement disrupts the energy comprising that frequency. Energy is absorbed or emitted to compensate for the disruption. Once a frequency changes, it maintains the new frequency until it is disrupted again.

As Einstein predicted, when *mass* increases in speed, it increases in energy, increasing its overall *mass*, while contracting in size. This is explained by a change in electromagnetic frequencies of the orbital patterns comprising *mass*. An increase in frequency requires more energy. More energy changes the frequency by decreasing the size of its wavelength. Thus, the absorption of energy increases the electromagnetic frequency, which increases the momentum of *mass* through space, increasing its *mass*, while contracting its size. This correlates exactly with the Lorentz transformation. This means that electromagnetic waves can absorb and emit energy. They change to a higher frequency with the absorption of energy or a lower frequency with the emission of energy. This then changes the size or shape of the orbital patterns that determines the speed *mass* moves through space. As the electromagnetic waves of orbital patterns absorb energy they decrease in size (wavelength) and increase in speed, and visa versa. As they emit energy, they increase in size (wavelength) and decrease in speed. This explains the contraction of the Michelson and Morley apparatus resulting in the contraction formulas by Lorentz and Fitzgerald.

Einstein knew that gravity would eventually be explained in conjunction with electromagnetic waves. This is exactly what we will do in the next chapter.

Quantum relativity explains how the same object—according to its speed through space—experiences differing amounts of energy within its orbital patterns. The same beanbag literally contains differing amounts of energy in different inertial frames. Had Galileo been able to observe the energy level of the orbital patterns of his new inertial frame of the moving ship, he would have noticed that they were indeed different than the energy level of the orbital patterns of the land. Each inertial frame produces a measurable difference in energy level of orbital patterns on a quantum level for the same object, causing it to experience a measurable increase or decrease in the overall energy from which it is comprised. The same *mass* can contain differing amounts of energy within its orbital patterns, creating differing inertial frames for the same *mass*.

This caused me to observe changes in energy. For instance, I noticed as my car accelerated, the ID card I wear at school but hangs on my mirror had swung at an angle opposite to the direction of acceleration. As the speed of the car evened out, the ID card had swung back towards its hanging position, making smaller and smaller back and forth movements until it was at rest within its new inertial frame. I knew that these back and forth movements were not a force from without, acting on the ID card, but rather, that the back and forth motions were changes happening on a quantum level until the ID card reached a state of equilibrium with the new inertial frame of the car to which it was attached. The reason it made back and forth movements after it had swung back towards its hanging position is because it acquired energy due to gravity. Its back and forth movements indicate a higher energy level then the car until the emission of energy put it at equilibrium with the energy level of the car.

Quantum Relativity is the result of differing energy levels of orbital patterns of *masses* in inertial momentum. Quantum relativity explains Special Relativity from a quantum perspective without the use of *time*. It explains the momentum of *mass* in different inertial frames. It explains how freely moving energies travel at the same speed as the *confined energies* of *mass*, maintaining Maxwell's constant. All of this sprang from a simple question for a fifth grade science fair experiment: *How can the same falling object travel two different distances?* The next chapter deals with the *acceleration of mass* caused by the absorption of energy—*quantum gravity*.

Chapter 8

Quantum Gravity

○ ○

Gravity is not a function of space-time, nor is it an attraction of masses, but rather, the source and cause of gravity is the acceleration of mass caused by the absorptions of energies originating at the quantum level.

Einstein's happiest thought was the realization that gravity is acceleration. He used an example of an elevator being pulled up in empty space at 32 feet per second per second. This would create the same gravitational field that a person experiences standing on the earth. The idea that gravity is acceleration begins the basis for quantum gravity.

Quantum gravity is the fourth leg to the Missing Model of Motion and is also based on the energy level of orbital patterns. The same principle that accounts for quantum movement accounts for quantum gravity—mainly that changes in the energy level of orbital patterns are changes in momentum. As the orbital patterns of *mass* absorb energy, the *mass* accelerates in the direction from which the energy is absorbed. Thus, gravity is nothing more than absorbed energies within *mass* initiating changes in the energy level of orbital patterns causing acceleration in the direction of absorption.

A unique phenomenon of *mass* is the relationship between energy levels and orbital patterns. Changes in orbital patterns initiate changes in energy levels and changes in energy levels initiate changes in orbital patterns. These simultaneous changes are a single process we discussed in the previous chapter and will discuss in greater detail in this chapter. In quantum gravity, the changes in the orbital patterns of *mass*, caused by an increase in energy, creates a movement of 32 feet per second per second—near the surface of the earth—in the direction from which the energy is absorbed. This is gravity—changes in the orbital patterns of confined energies initiated by changes in their energy level.

An example of the effects of gravity is a penny falling off the Space Needle in Seattle. As the orbital patterns of the penny absorb energy emanating from the surface of the earth, its orbital patterns simultaneously change, increasing its speed in the direction of absorption 32 feet per second per second until it reaches terminal velocity. At terminal velocity, the acceleration caused by gravity—absorbed energy causing changed orbital patterns—reaches equilibrium with the forces in the air that are pushing against it from a different energy level. This interaction creates a steady momentum for the falling penny—the orbital patterns are remaining constant and are not changing. At this point, energy is absorbed and emitted at the same rate. When the penny hits the ground, its increased energy level interacts with the lower energy level of the ground. Energy is emitted and transferred into surrounding matter until the orbital patterns of the penny and the surrounding matter reach equilibrium with the orbital patterns of the earth. During this whole process, the conservation of energy in the universe is unchanged.

When the penny hits the ground, there is an immediate change in the orbital patterns of the penny. This change initiates a simultaneous change in the energy level of the penny. Energy is emitted from the penny and absorbed into nearby masses or travel freely out into space. Orbital patterns of the substances the penny hit or its emitted energies affected through absorption, such as the pavement and the surrounding air, were also disrupted, causing the absorption and emissions of energies according to their equilibrium needs. The energy level of orbital patterns change until equilibrium is reached for the *masses* involved and the energy level of orbital patterns obtain equilibrium with the energy level of the orbital patterns of the earth. The energy level of the orbital patterns of the penny remain synchronized with the energy level of the orbital patterns of the earth until another disruption occurs such as when somebody picks up the penny. The penny stays at rest on the earth's surface because it is continuously absorbing and emitting energy at an even rate due to the terminal velocity reached because of its continuous contact with the surface of the earth. The energy necessary for acceleration is continuously being absorbed into the penny. Because its orbital patterns cannot change due to its terminal velocity, they are emitted at the same rate of absorption. Remember, this all takes place at the speed of light. Thus, gravity is the absorption of energy, causing a simultaneous change in the orbital patterns of *mass*, accelerating the *mass* in the direction from which the energy is absorbed.

This is the essence of gravity. This is why Einstein was unable to unify general relativity with quantum mechanics. The necessary bridge between macrophysics and microphysics— the Missing Model of Motion—was not yet discovered. Einstein didn't consider the cause or source for the momentum of *mass* through space from a quantum perspective. Instead, he brilliantly created space-time to describe gravity on a large or macrophysics scale. Thus, he accurately predicts the effects of gravity, but fails to explain how gravity operates from a quantum perspective. Space-time describes distortions in space caused by large masses such as suns and planets and the effect this has on moving objects near their surfaces such as other planets, moons, asteroids, and light. In reality, these distortions in space are actually higher concentrations of energy near the surfaces of large spherical masses such as the sun or earth. This will be explained momentarily.

As you can see from the explanation of quantum gravity, gravity is not a mysterious force of attraction between *masses*, nor is it the effects of warping space-time, but rather, it is forces acting within matter. It is the effects of absorbed energy on already moving *mass*, meaning the momentum of *mass* through space. Quantum gravity is only one part of a four-legged stool. The other three descriptive parts are quantum momentum, quantum movement, and quantum relativity. These four parts comprise the Missing Model of Motion. The Missing Model of Motion explains why objects move in the first place. Without understanding why objects move in the first place— the source or cause of momentum—one cannot understand how *mass* is accelerated. Then by adding an understanding of quantum movement to quantum momentum, gravity can easily be explained as acceleration caused by changing orbital patterns initiated by the absorption of energy, causing *mass* to accelerate in the direction from which the energy is absorbed.

How does the earth sustain the constant flow of energy emanating from its surface in order to maintain the constant effects of gravity? The earth is already in momentum around the sun by the orbital patterns from which it is constituted. The amount of energy it receives from the sun allows it to continuously accelerate around the sun without accelerating into it. As energy is absorbed into the spherical *mass*, *mass* either accelerates or—like the penny reaching terminal velocity—acceleration is restricted and energy is absorbed and emitted at the same rate. Eventually, the constant flow of energy being absorbed and emitted by *masses* that comprise the earth has nowhere to escape except through the surface, and when it reaches the surface, it is then emitted into space.

Illustration—10

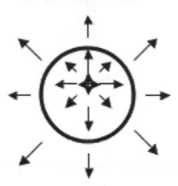

Due to terminal velocities restricting the acceleration of spherical *mass*, absorbed gravitational energies are emitted towards its outer surface, creating concentrated amounts of energies near the surface.

Emitted energy from any *mass* causes other *masses* around it to accelerate towards the first *mass*. This continuous process of *mass* emitting energy causing other *masses* to accelerate towards it produces the spherical shapes that we see in the form of suns and planets. The absorbed and emitted energies eventually have nowhere to go or escape except through the surface of the sphere, moving out into space. This causes any object on or near the surface to accelerate towards the surface—gravity. The effects will be strongest closest to the surface and will get gradually weaker as energy spreads out the further it gets away from the spherical surface. Our earth and other planets in our solar system are beneficiaries of large amounts of this energy from the sun. Our sun receives this energy from all sources that absorb and emit such energy such as stars in our galaxy, the center of our galaxy, and all the other galaxies in the universe. It is a reciprocal process that keeps energy moving throughout the universe. As you visualize this process of massive spherical bodies continuously absorbing energy and emitting unused portions through its surface areas into the immensity of space, remember that it all takes place at the speed of light.

Physicists before Einstein's quanta theory of electromagnetic waves thought there was a substance that filled the immensity of space called ether through which light propagated. Interestingly enough, the universe is actually filled with energy moving at the speed of light, which is being absorbed and emitted through a reciprocal process; it just doesn't need a medium to move through space. Some areas of space have higher concentrations of this energy such as near the surfaces of large massive spherical bodies like the sun and earth. The high concentration of energy near the surfaces of spherical bodies explains Einstein's use of space-time to predict the movement of masses or traveling light near the surfaces of large, massive, spherical bodies. The distortions described by space-time are actually concentrated amounts of energy in comparison to the less concentrated amounts of energy in open space. The surface area of any spherical body will emanate the largest concentrations of energy before they dissipate as they spread out in open space the further they get from the surface.

This theory supports Einstein's theory for light as quanta. As mentioned earlier with the example of the rope, the Missing Model of Motion does not need a medium for freely moving energy to travel through, such as ether. When you drop a stone in water, it appears that energy moves through the water in the form of waves. What is really happening is the absorbed energy that caused the stone to accelerate disrupted the orbital patterns of the lower energy level of the water. The waves in the water are the chain reactions of disrupted orbital patterns and their subsequent absorption and emission of energies to compensate for the changes. The waves are the results of changes in momentum and movement of the water molecules on the quantum level. The energy didn't need water to travel through. The energy was the water at varying different energy levels, causing temporary shifts in momentum and movement on a quantum level. This created the waves. What we see in the slow moving waves are transactions that are happening at the speed of light.

The absorbed energy due to the effects of gravity is visible. When I hold a pencil several feet off the ground and drop it to the ground, the absorbed energy responsible for the orbital changes and acceleration in movement causes the pencil to bounce a few times before that energy is emitted and the orbital patterns of the pencil change to coincide with the orbital patterns of the earth with which it now shares an equilibrium state. When my car comes to a quick stop, the nametag hanging from my mirror continues to swing back and forth until the energy level of orbital patterns reaches equilibrium with the new energy level of the car of which it is attached to by a string. It slowly emits energy as orbital patterns shift as the nametag collides with lower level air particles in the space surrounding it.

When a ball bounces, it wants to go in the same direction like the water container that is pushed by an astronaut in the free-falling space shuttle. The orbital patterns of the ball are the movement and momentum of its travel. As the ball hits the ground, the extra energy comprising the orbital patterns shift and the momentum of the ball changes, causing it to move away from the surface of the earth. It would continue in this direction uninterrupted if it weren't for energy being absorbed due to the effects of quantum gravity. As the ball is heading in the direction of its new momentum, orbital patterns change as energy is absorbed into them due to the effects of quantum gravity. We watch the ball reach its apex, and then it begins to accelerate towards earth. What we see as a slow process is actually playing out at the speed of light. For a brief moment, at its apex, the energy level of orbital patterns of the ball was equivalent with the earth. With each time the ball bounces, orbital patterns are disturbed and changed and energy emitted until eventually the ball comes to a resting position on the earth. When the ball looks likes it is resting on the earth, it is still moving through space. Its energy level of orbital patterns is still the source and cause of its movement through space. It is just connected to the earth due to the energy emanating from the earth and the effects of quantum gravity. If I shut off the energy emanating from the earth and threw the ball into the air, it would move away from the earth like a helium balloon escaping from the grasp of a child. In the case of the ball, its energy level of orbital patterns would be the source and cause of its movement through space.

What comes first, orbital changes or changes in energy levels? As we mentioned earlier, a change in orbital patterns initiates a change in the energy level, and a change in the energy level initiates a change in the orbital patterns. It is like blowing air into a balloon. As the balloon receives more air, the boundaries of the balloon expand. As the balloon loses air, its boundaries contract. They are part of the same process. This is due to frequency changes of the electromagnetic waves that comprise the orbital patterns.

As energy is absorbed on the atomic level, electromagnetic energy shifts to a higher frequency. This shift changes its corresponding wavelength. As energy is absorbed, the wavelength decreases in size as the frequency increases in rate. This creates a contraction of the *mass* in the direction of movement while increasing its momentum through space as predicted by Special Relativity. The reason *mass* accelerates in the direction of absorption is because that is where the change of frequency is initiated, causing the *mass* to accelerate. The orbital patterns shift at the point of orbit where the energy is absorbed, causing an acceleration in that direction.

The quantum cause of gravity is energy just as Einstein predicted. Energy travels at the speed of light and is part of the same energy by which all *mass* is comprised. Energy is continuously absorbed and emitted by all *mass*. When orbital patterns reach terminal velocity, the energy is transmitted through *mass* by continuous absorptions and emissions. The energy level of orbital patterns is a fundamental part of the structure of all atoms and the reason why atoms move through space in the first place. The energy level of orbital patterns is why *mass* can speed up and slow down as it move through space. It is also the cause for the expansion and contraction of *mass* as it speeds up or slows down. *It is the cause of the acceleration we call gravity.*

Gravity is initiated through absorbed energy. The results of its effects are observable and measurable. As energy increases in accelerating *mass* towards the surface of the earth or as energy decreases when the accelerating *mass* comes to an abrupt stop, the effects and results are observable and measurable. When I take my ring off my finger and let it go a few feet above my laminate flooring, energy is absorbed into the ring, causing it to accelerate towards the surface of the earth. When the ring hits the floor, it is at a higher energy level than the floor. The ring bounces and moves until its energy level reaches equilibrium with the energy level of the laminate floor, (earth). There, it will remain at the same energy level or inertial frame as the floor, (earth), until its energy level of orbital patterns is disrupted again. When I held the ring above the floor, it is said to have potential energy. Potential energy is nothing more then potential changes in orbital patterns—acceleration—before a terminal velocity is reestablished.

Quantum gravity explains the moon and tides. As the moon absorbs and emits energy as any spherical body does, the emitted energy travels towards the earth causing changes in the energy level of orbital patterns of all *mass* that absorbs it. The greatest impact would be near the surface of the earth facing the moon as the energy level of orbital patterns is changed in the direction of their source, causing the ocean waters to move towards the moon until it reaches terminal velocity due to energy emanating from the earth accelerating the water in the opposite direction. This is an ongoing process transpiring at the speed of light. The moon doesn't mysteriously attract the water or warped space causing the water to move towards the moon, but rather, it is energy emanating from the large, spherical-shaped moon accelerating the *mass* of the water towards the direction of absorption. I imagine it is the magnetic, fluid qualities of H_2O that accentuate the effects, elongating the water towards the moon, causing the phenomenon of tides.

This theory also explains the orbiting of planets and satellites. The momentum of the orbiting planet or satellite is maintained at a certain velocity due to quantum momentum. The energy emanating from the satellite's source alters the satellite's energy level of orbital patterns enough in the direction of its source to create an orbital or elliptical path around the source just as we have been taught in basic physics. Its momentum and increased acceleration in the direction of the source of absorbed energy are the reasons it maintains its constant orbit. Too much acceleration will move the satellite into its source, and too little acceleration will cause the satellite to slowly drift away from the source. An extremely large asteroid could collide into a satellite such as the moon to increase or decrease its quantum momentum and cause the effects of coming closer or drifting further from its source. *The energy level of orbital patterns of confined energies maintains the momentum of mass through space. Absorbed energy, changing the energy level of orbital patterns of mass, causing acceleration in the direction of absorption, is the source and cause of gravity.*

Einstein's great epiphany that gravity is acceleration completes the Missing Model of Motion with quantum gravity being the final leg. Thus, quantum momentum, quantum movement, quantum relativity, and quantum gravity complete the Model of Motion and accounts for the momentum and *movement* of all *mass* through the universe—from a quantum perspective.

Newtonian physics is based on quantum interactions explained in the last four chapters—the same quantum interactions that govern the universe. Newton's most impressive observation, in my opinion, is his first law that basically states that an object in motion tends to stay in motion until it is acted upon by an unbalanced force. This law from a quantum perspective is the basis of quantum momentum, quantum movement, quantum relativity, and quantum gravity. From this law springs the quantum explanation for Special and General Relativities without the use of *time*. All momentum and movement is initiated and maintained on a quantum level.

The four legs of the Missing Model of Motion explains many phenomenon from a quantum perspective such as why two masses are attracted to each other, the source and cause of momentum and *movement*, tides, high volumes of energy near the surface of massive-spherical bodies, and why the effects of gravity are lessened the further you get away from the surface. In the end, our purpose is not to compose reality to match our perceptions but rather to change our perceptions until they conform to reality—to see things as they *are* even if it is different from how we want them to be. This is why we ask the next question. *Does time really exist?*

Chapter 9

The Illusion of Time

○ ○

Time is sand passing through an hourglass.

Tick…Tick…Tick…Tick…

We've all heard it…when sleep fails us…and we lay awake in the wee hours of the morning listening to its incessant beat. There is never any drum-tick, jazzy cadence…no rhythmic pause here and there…just the never-ending, never-deviating pulse. In the darkness…in the light…it's always there.

Tick…Tick…Tick…Tick…

We place them on our bedside tables…we hang them on our walls… we wear them—these sentinels of existence. They're small and simple in design and function, but magnanimously complex in context and theory. We use them to measure…to plan…even to dictate the day-to-day events that weave the fabric of our lives. Our activities and responsibilities—individually or collectively, locally or globally—hold inferior position to these small, mechanical wonders…clocks!

But, what do clocks do? Sounds like an easy question to answer. We all know what it is…right? It's something we've known since our earliest recollections. We're taught about it from our first weeks of school. We learn to tell it. We're constantly asked about it from passers-by on the street. We watch *it* pass…we squander *it*…we track *it*…we lose *it*…we covet *it*…and we borrow *it*. But, what is *it*…really?

Time.

Aside from the theory of simple forces applied in mechanical movement, is there an actual force called 'time' that moves the hands of the clock? Does it change the numbers on your digital watch...push the earth...or drive the moon? Does it instigate the season's change or cause our cells to degenerate and our memories fade? Does it exist outside the consciousness that defined it...that created it? Is it a physical entity...with shape and form...matter with power to act, and thereby, be acted upon?

No.

Then what is time? In the tangible realm of general physics—the non-quantum world—no equation or computation can stand without it. "Distance over Time" (d/s) permeates all theories of macrophysics. No ball can be thrown, no cannon shot, no train can travel, no rocket launch without including 'time' as a parameter of the solution to any myriad of associated story problems or experiments. It is oftentimes referred to as the fourth dimension, having an equal part in the continuum of space. It is something that simply...*is*.

Or, *is* it?

A study of quantum mechanics, even infantile in depth and breadth, would argue to the contrary. Unlike general or astrophysics, quantum theories abhor the concept of time. In this realm of science...the "sub-atomic" view of the universe...time does not exist. There is no precedence for d/s...no formula or equation containing time as a variable. It is irrelevant. Some may contend as to the express details in reasoning, but, simply put, it is a matter of our inability to gauge with precise clarity the exact location in space of an atom's electron at any given moment in 'time' that compounds the issue. But, does our lack of instruments accurate enough to measure individual electrons and their movements about the nuclei of atoms necessarily corroborate a stand against the existence of time? Our feeling is that our technological accomplishments and creations are irrelevant to the real issue—that of 'time' itself being a purely man-made creation.

Thus, time is a product of human consciousness and serves solely as a humanly intrinsic reference point for the linear measuring of these experiences. It is an arbitrary concept (consisting of limitless range and scope) that attempts to quantify in finite terms...the "infinite"—that very thing which has neither beginning nor end—eternity. 'Time' is indefinable as an absolute entity of the cosmos—holding neither place in space nor exclusive power to change. Take, for example, a brief comparison between the idea of an "inch" and a "second." Both of these units of measurement are simple figments of human imagination. Neither one exists separately from the consciousness that creates and defines them. "Inches" and "seconds" are both arbitrary units of measurement created, defined, and generally accepted by the majority of society to exist and each respectively defines a precise amount of something. These definitions, however, are only inherent to the social/ scientific consciousness that obliges them. In other words, they are only valid *here*—on planet earth. There is no universal decree of human preeminence that places our feeble and infantile level of comprehension superior to that of other possible forms of intelligent life…let alone—above that of God.

At this point, one may argue this to be a simple case of semantics… "inches, feet, meters…seconds, hours, days"…what's the difference? They are all units to measure something. Yet, "inches" and "seconds" are finite attempts to capture the infinite. "Inches" attempts to capture a portion of distance, which is a tangible component of space, while "seconds" attempts to capture movement, which is a tangible component of energy. But do "inches" and "seconds" exist outside of conscious thought? Absolutely not! Distance and movement exist outside of conscious thought. But "inches" and "seconds" are human attempts to quantify distance and movement.

So then, what is our definition of time? Time is a measurement of contrasting movements. It is ultimately a measurement of energy, that which cannot be created nor destroyed. In our world, time is the measurement of repeating patterns such as the earth rotating (days) and revolving (years). These patterns become the basis by which all other movements are compared. Then these patterns are divided into smaller but equal parts. We duplicate these natural existing patterns and their divisible portions through mechanical means such as the astutely designed Swiss watch or the Japanese digital wonder.

In the end, there is nothing infinite about the concept of time. It is an arbitrary attempt to measure and cognitively capture contrasting movements. Although one could pose the argument, given society's current paradigms, that a 'second' segments or defines a precise amount of time, the proposal that one could then in turn segment something as broad and infinite in reality as eternity with something as inert and arbitrary as time—having neither form, nor power, nor place in space—is absurd. 'Time' is contrived from human consciousness to define and give order to man's being and the passage of experiences. It serves solely as a reference point—an anchor—for the mortal, linear-thinking mind of man. It does not coexist with the omnipresent state of matter in the universe—matter that can be neither created nor destroyed . . . matter that is eternal in its existence. Time only exists so long as man's

consciousness depends on it. In perspective, it is difficult to imagine God...
the Alpha and Omega...the Almighty...ruling from His courts on high...
immortal...perfect in being and comprehension...possessing absolute and
unfathomable light and knowledge...eternal in His existence—being the
same yesterday, today, and forever...seated upon His Throne of Glory...
wearing a Rolex. Yet, in a way, our solar system is like God's watch. It moves
in an orderly mechanical pattern that repeats itself over and over until, like a
Timex, it eventually stops ticking.

If consciousness ceased to exist, so would the very concept of time with
its attendant quantifying units of seconds, minutes, years, and millennia. The
ideas of millimeters, inches, yards, and miles would cease as well. Space and
energy are not dependant upon consciousness for its reality and endurance.
Matter would still move, change, and occupy a precise and finite amount of
'space'—independent of the will of consciousness. Any statement regarding
"time" or "linear measurement", however, used to define the existence of that
same matter (in its various states) would serve as a purely arbitrary reference
point relative only to the consciousness that is seeking to quantify or define it
and nothing more. Matter is infinite it its scope of existence—its "being"—
regardless of its state. A "finite" characteristic of matter is distance—the
amount of space the matter occupies or the space separating two distinct
objects of matter. The 'time' of matter's existence, regardless of its particular
form, is nullified by the reality of its eternal nature. Any attempt to describe
or measure the existence of matter using 'time' is futile beyond that of a solely
human-based need for order and reckoning.

We must shed the linear paradigm of time with its accompanying
limitations of a '"past, present, and future" if we are to truly understand
the eternal, omnipresent state of matter and energy and its association with
eternity. There was no 'time' at which matter and energy, whose existence we
wish to define and quantify in the terms of 'seconds', 'minutes', or 'millennia'
did not exist. Nor is there any future point at which it will cease to exist.
Matter and energy are timeless. Likewise, there is neither beginning nor
end to eternity. Thereby, any attempt to justify and validate the existence
of a particular 'piece' or component separately from that which has neither
a point to begin nor point to end is meaningless outside the indigenous
parameters of the particular consciousness that has need to reference it. Time
is an indignity to the essence and intelligence of God. Things simply "are"
in terms of existence in the omnipresent consciousness of God. It is one
eternal round, having neither beginning nor end—no past nor future—just
the present. In other words, 'time' does not exist outside the context of our
mortal existence.

Chapter 10

Timelessness

○ ○

Timelessness is the absence of conscious thought.

In the previous chapter, we expressed our view that time, as we measure it, doesn't exist separately form conscious thought. It is a byproduct of conscious thought. The opposite of time is *timelessness*. *Timelessness* is the foundation upon which our theory, *The Omnipresent State of Mass-Energy and Timelessness*, is based.

So let's get into our theory and explain what we mean by *timelessness*.

It starts out with our idea of omnipresence. Normally omnipresence means that someone or something is present everywhere. That is a difficult concept to grasp, trying to imagine something simultaneously everywhere in the same instance. It would have to be all encompassing such as the idea of space. But for us, the word takes on a different meaning. For us, omnipresent means *always present* in reference to the concepts of past, present, and future. In other words, always existing in the present state, never deviating into a past or a future state. Omnipresence describes the state of physics as expressed in this book, and it describes spirituality, which will be expressed in another book.

It is very interesting that at least two philosophers have pondered the concept of *timelessness*, Parmenides (515-450 B.C.) and Augustine (354-430).

Here is what we found in an introduction to philosophy book on Parmenides, a philosopher from southern Italy. It reads:

By employing strict logical argument he produced an interesting idea about Time: all that actually exists is the immediate present. Talk about the past and the future is just talk—neither has any real existence. (Robinson and Groves 1999)

Here is an interpretation of Augustine's struggle to locate his *self* in *time*:

> During the last three years of the fourth century in the Roman city of Hippo in North Africa, a middle-aged man, Augustine, wrote of his life crisis. Among other things, Augustine seems at one point to have lost his 'self.' He could not locate his 'self' anywhere in 'time.' This distressed him. Augustine reasoned thus: The past does not exist—it has left only footprints. Similarly the future does not exist—for it is not yet. But the present also does not exist in a way that one can grab hold of it; it has no 'extension' or 'duration.' By the time one pronounces the word 'present,' the first syllable is forever gone. Hence the crisis: If the past does not exist except in memory, if the future does not exist except by anticipation, and if the present does not endure: Where am I? When am I? (Barlow 2007, 1)

What both of these thinkers point out is that it is logical to question the nature of time and the actual existence of a past and a future separate from the present. The second thinker even questioned the existence of the present because it had no endurance or duration from one moment to the next. You cannot grab a hold of it. We want to explain why we think these thinkers were on their way to understanding the true nature of our existence. This is where we introduce the first part of our theory, *the omnipresent state of mass-energy*.

The law of conservation of mass-energy used to be two separate laws up until Einstein. It was the conservation of matter and the conservation of energy. Now it is one law. It basically states that matter and energy cannot be created nor destroyed but can change forms, such as matter becoming energy and energy becoming matter. What this essentially says is that you are not going to add to or take away from what already exists.

But how did matter and energy come into existence? If matter and energy cannot be created nor destroyed, it would thus be reasonable to think that its existence had no beginning nor will it have an end. That thought is the concept of eternal, no beginning or ending. No matter how far you look in either direction, within the idea of past or the idea of future, matter and energy cease not to exist. Matter and energy are always present, or omnipresent, according to our new definition for that word.

The conservation of matter and energy suggests that anything that presently exists has always existed in some form or state, from eternity to eternity, from no beginning to no ending. This idea goes beyond the ability of human consciousness to grasp. Our present existence is built on the idea of a beginning and an ending, a birth and a death. Everything about us seems to have a beginning and an ending, such as life, civilizations, planets, solar systems, and even galaxies. Yet, the fundamental substances that make up and allow for the existence of those mentioned above are eternal in nature, meaning they have no beginning or ending. They have always existed and will always exist. In essence, they have no *time*.

This is what we mean by the omnipresent state of mass-energy. The fundamental state of all that exists is without a beginning and an ending. It exists in a perpetual, non-static state. We refer to this state as omnipresent, or omni-existing. You cannot add to or take away from the days of its existence. It is always present, thus, omnipresent, or *The Omnipresent State of Mass-Energy.*

Timelessness describes the state of matter and energy.

Timelessness, the second part of our theory, is used to finish the definition of the first part of our theory: the omnipresent state of mass-energy. Not only is mass-energy omnipresent, it is also *timeless*. As a description, it suggests that *time* is not a physical force separate from or woven into mass-energy. As our missing model of motion suggests, mass is confined energy. Mass does not exist separate from energy, nor energy from mass. The fundamental state of mass-energy is *timeless* and as stated in modern physics, it cannot be created nor destroyed.

Time is the incremental measurement of the movement of energy, whether confined as mass or moving freely through space. Without energy—*movement*—there is no need for the concept of time. Instead, time developed as a conscious tool for sequencing movement, giving it a pretentious beginning, middle, and end.

Timelessness yields the reality that there is nothing that exists in its most basic form, whatever that may be, that hasn't always existed or will cease existing. Because of this, *timelessness* yields a few realistically logical consequences. First of all, with *timelessness*, the finite is just the organization or reorganization of the infinite, and second, the past and future do not have a separate existence from the present. These are topics of subsequent chapters. And third, the reality of timelessness focuses spirituality towards *omnipresent-living*, the main topic of our next book.

Earlier, we mentioned two philosophers who couldn't mentally conceive of an existing past and future that is separate from the present. What about modern thought? Have any recent philosophers or physicists entertained the idea of timelessness?

A couple of years after we formulated our theory, David pointed out an article he had found in Discover magazine. Part of the caption under the title could have been written by us. It read:

Imagine a universe with no past or future, where time is an illusion...
(Folger 2000, 54)

We couldn't wait to read and discuss this article.

When we read it, we realized that Julian Barbour's ideas of timelessness were completely separate and unrelated to the theory we worked out. Yet, the article did say a few things that really caught our attention.

Julian Barbour was quoted as saying:

> Given what a fascinating thing time is, it's surprising how few physicists have made a serious attempt to study time and say exactly what it is. (Folger 2000, 57)

Also in the article, Don Page, a cosmologist at the university of Alberta in Edmonton is quoted as saying:

> I think Julian's work clears up a lot of misconceptions. Physicists might not need time as much as we might have thought before. He is really questioning the basic nature of time, its nonexistence. (Folger 2000, 61)

We bring up this article that introduces Julian Barbour's book *The End of Time*, to show that the possibility of *timelessness* is still being considered today.

Yet, the possibilities of *timelessness* as theorized by us is different than anything we have yet to encounter.

Our theory coincides with the law of conservation that states that mass-energy cannot be created nor destroyed. This law helps explain what we mean by *omnipresence*. It states that mass-energy, in its most basic state, wasn't created. It didn't come from nothing or a non-existent state. It also states that it will not one day cease to exit or stop existing, just fade into nothing. Although its present form is in a constant state of change, its perpetual existence is never in question.

This is the big paradigm shift. The most basic substances, particles, or energies of the universe have always existed and will always exist. They are infinite in duration. How they exist, how they are bonded, or how they interact is in a constant state of change. The universe is never the same universe from one moment to the next, yet the substances, particles, and energies of the universe are constantly present, or *omnipresent*, in that they are *timeless*. This is what we mean by the omnipresent state of mass-energy and timelessness. The paradigm shift is in realizing that everything about us, including the stuff that makes up us, is eternal, ever-existing, without beginning of days or end of years. It is forever. There is nothing about us that hasn't always existed and will not always exist. It is the basis for understanding the difference between infinite and finite. It is also the basis for understanding that the past and future do not exist separately from the present. There is no separate past or future with revolving doors to the Here and Now.

The next two chapters deal with two interesting consequences of *timelessness*. Chapter 10 differentiates between the infinite and the finite. Chapter 11 explains how the past and future are not separate from the present.

Chapter 11

Infinite and Finite

○ ○

That which has no beginning has no end.

Every theory has its logical consequences. One of the logical consequences of *The Omnipresent State of Mass-Energy and Timelessness* is a clear-cut distinction between the infinite and the finite. In short, the finite is the temporary arrangement of the infinite.

Let us first define what we mean by infinite. The infinite is the indivisible core of mass-energy that cannot be created nor destroyed. No matter how you arrange it, stack it, bond it, or break it up again, it maintains its perpetual existence. It exists in and of itself, *independent of any other factor*. As of yet, science is still attempting to determine the smallest units of what makes up the omnipresent state of mass-energy that we presently interact with on a moment-to-moment basis. Whatever we finally determine this to be, it will be the infinite from which all things spring. And by being used as building blocks, it ushers in the finite, which is organized from the infinite.

Our definition of the finite is the arrangement and disarrangement of the *timeless* infinite. The finite has a dependent existence. It is dependent on the present ordering of the infinite.

The ideas of the infinite and the finite go hand in hand in explaining all that exists. For example, at one point, the earth as we now know it did not exist, yet the materials that makes up the earth have always existed. At some point, the earth as we now know it will not exist, yet the materials that presently make it up will continue in their perpetual, *timeless* existence. The earth was organized or came together from timeless mass-energy, and when it is disorganized or comes apart, the timeless mass-energy from which

61

it was made will continue to exist. In this sense, the earth as a spherical shape is considered finite while the mass-energy that presently makes it up is considered infinite.

An analogy would be a set of legos. You can take a set of legos and arrange the individual pieces into countless configurations by putting them together and taking them apart. The individual legos themselves represent the infinite, while the different configurations of how they are arranged represent the finite.

At this point, we would like to introduce consciousness from an infinite and finite perspective. Consciousness is the greatest gift we enjoy on a moment-to-moment basis. Without consciousness, I am not; you are not; we are not. Einstein once said something to the effect—when referring to our ability to understand the basic laws of nature—that God was not malicious. Could there be anything more malicious then the annihilation of consciousness at death's doorway? For this reason, we believe consciousness has its roots in the infinite.

Our present state of consciousness is dependent on the *finite* organization of the infinite. As far as we can tangibly prove at this time, there is nothing about man, including his thoughts and ideas, that is not dependent on the present organization of mass-energy from which he is presently constituted. Even the elements that align to form thoughts and ideas are finite arrangements of the infinite because they depend on another factor for their existence, the present arrangement of mass-energy. If you disorganize the arrangements that maintain them, their existence ceases. An example is the idea of an inch, which we brought up earlier. An inch is an idea of consciousness. It is used as a tool to measure the length of mass. The idea of an inch doesn't exist separate from thought. It is the product of thought. If you take away conscious thought, (disorganize it), the idea of an inch ceases to exist. Yet the mass-energy that made up the thought of an inch continues to exist in its omnipresent state.

Now we can use *infinite* and *finite* to explain the existence of time. Time is a *finite* idea to measure movement from point to point or event to event. Like the idea of an inch, words such as second, minute, hour and year are *finite* ideas dependent on another factor for their existence—consciousness. Their meanings are derived and maintained as appendages of consciousness. Time, as a finite idea, gives consciousness a reference point from events that have already happened to events yet to happen. From this point of reference, we can analyze, measure, reminisce, hope, plan, and predict. Yet, *time*, as an appendage of consciousness to measure movement or order events, is a *finite* idea that depends on another factor for its perpetual existence. In this sense, it is no different then the idea of an inch.

It is the *finite* idea of time that creates linear thought because our point of reference is *Now*. Thus we reason that anything prior to *now* is the past and anything following *now* is the future. The problem with this way of thinking is how does this *now*, (a few moments later), differ from this *now*? The amount of mass-energy in the universe hasn't increased or decreased. What we now call the past is a *finite* thought of an event prior to our present reference. The past and the future are always referenced from our present point of *now*. This point of reference is the timeless state of mass-energy from which consciousness flows. Yet, conscious thought gives an illusion of separation from the omnipresent state of mass-energy from which yields its existence. This separation creates linear thought and the illusion of things appearing and disappearing in space and time. But in the reality of mass-energy, there is no past or future, there just "is".

This is where we reemphasize the paradigm shift from *time* to *timelessness* and from *finite* to *infinite*. Omnipresent mass-energy is *timeless* and *infinite*. It exists in an *eternal now* only changing in how it is organized and arranged. *Time* is not a part of the structure of omnipresent mass-energy, except in the sense of duration.

Time is to *Finite* as *Timelessness* is to *Infinite*.

Chapter 12

The Past and Future

There is no forward or rewind button in the universe.

Another logical consequence for the *Omnipresent State of Mass-Energy and Timelessness* is that the *past*, present, and *future* exist simultaneously.

When we think of the past, we are referring to events that have already happened, and when we refer to the future we are projecting events yet to come. *Timelessness* implies that the *physical* existence of each is not separate from the present. How is this possible? It is possible in that all the mass-energy that made up any past event presently exists today. And all the mass-energy that will make up any future event presently exists today, too. There is no mass-energy of the past that doesn't presently exist, and there will be no mass-energy in the future that doesn't presently exist. From an omnipresent state of mass-energy and timelessness perspective, the past, present and future exists simultaneously. The only differences between the *past*, *present*, and *future* are the finite arrangements of the infinite.

A *past* of a previous finite arrangement does not and cannot exist outside of *omnipresent* mass-energy. A *future* with its potential finite arrangements does not and cannot exist outside of *omnipresent* mass-energy. And even though *omnipresent* mass-energy is continuously rearranging, giving the appearance of changing events, its perpetual existence isn't added to (created) or taken away from (destroyed)—the conservation of mass-energy.

Imagine being told that the past doesn't exist? How would you react to such a claim? When we first encountered the possibility of the nonexistence of time, we realized one very important consequence. If time didn't exist, that meant the past and the future didn't exist separate from the present, except in the cognitive structures of the brain driven by consciousness. This was a deep thought. I shared this idea with a friend. When I told him the past doesn't exist, he seemed appalled.

"What!"

"The past doesn't exist!" I told him again. He shook his head in disbelief as we entered into a fun debate as my wonderful wife tried to be the mediator of our different perspectives.

I didn't claim that the past didn't happen. I made the statement that the past, as we understand it, doesn't presently exist. This is one of the consequences of timelessness.

A paradigm shift is needed to experience the *past* and *future* as part of the present.

The idea of *past* and *future* is a creation within consciousness contained in present memories. Consciousness gives us the ability to remember prior arrangements of mass-energy or project future arrangements. The ideas of a *past* and a *future* are appendage to the idea of *time*, creating a before and an after—but before and after what?

Now!

Existence is *Now*. And then, *Now*. And then, *Now*. But what is this *Now*? It is the omnipresent state of mass-energy from which consciousness flows. Without consciousness and memory, the ideas of *time*, the *past*, and the *future* do not maintain an existence separate from *now*. Time, past, and future are concepts available to consciousness via the brain to capture, create, maintain, and experience events, hold them in sequential order, and plan and anticipate new ones. Yet all of this continuously takes place within the flow of *Now*.

The omnipresent state of mass-energy and timelessness is the *Nowness* through which consciousness flows. The total-mass energy of this *Now* is the same total mass-energy of this *Now*, (one second later). The total amount of mass-energy has not changed—just its arrangements. It is from the *Now* that we remember past arrangements. It is from the *Now* that we project future arrangements, but we never leave the flow of *Now*. The mass-energy of any past arrangement is the same mass-energy of any present arrangement, which is the same mass-energy of any future arrangement. The *past*, *present*, and *future* are finite arrangements of omnipresently existing mass-energy.

Here is an example to illustrate how the *past* doesn't exist separately from the present, showing how there is nothing from this *Now* that is left behind as I move into another *Now*. Wanting to remember the past, I take a picture of my three-year-old daughter with her curly red hair, who is looking like the next Broadway Annie. Five years later, I pull out the picture. When I look at the picture, I say that this event took place in the past (as if the past has a separate existence from the present and future). Yet, when I examine all aspects carefully, does the picture really represent a separately existing past that I could go visit if my time machine was working? Or, is the same mass-energy that made up everything represented in the picture still existing presently. There isn't any mass-energy that presently existed when I took the picture that doesn't presently exist five years later when I look at the same picture. None of the *mass-energy* was left behind at the moment the picture was taken five years previously. That is why time travel doesn't exist. There is nothing to go back to. It is all with us *Now*. The only differences in the *mass-energy* are the finite arrangements. Even the picture is comprised of omnipresently existing *mass-energy* that is maintaining its present finite arrangement. If the picture were to get burned in a fire, the *mass-energy* would rearrange but the overall *mass-energy* in the universe would remain conserved. In the end, the same *mass-energy* that existed then, exists *Now*, and will continue to exist forever. Timelessness implies endless duration of the infinite and the ever-changing arrangement of the finite.

Notice how *past* and *future* directly ties into the idea of *infinite* and *finite*. The *infinite* isn't added to or taken away from on a moment-to-moment basis, but how it is organized or exists in relation to itself is perpetually changing. It is this perpetually changing organization and relation that gives the notion of a *past* and *future*. Without consciousness and the ability to think, remember and predict, there is no *past* to hold on to or *future* to project. The *past* and *future* are just *finite* arrangements of the *infinite*.

Let's take an artifact from the *past*, say a Van Gogh painting. My conscious thought (which always takes place *Now*), references that painting (the moment it was painted) as prior to the *Now* I am presently experiencing. In reality, the finite arrangement that makes up the Van Gogh painting is still organized in the same manner as when it was originally painted. The painting doesn't also physically exist in some past, for everything that made up the painting then is with us *Now*. If it were to get destroyed by fire, the present finite organization would be disorganized, but the infinite, in its omnipresent state, would continue in its perpetual existence.

One day, that same friend that I had a discussion with about the *past* not existing left me a note that said, "I was not here because the past does not exist." With much laughter, I called him up and corrected his thought by saying, "You are still not here because the past does not exist." By the way, I still have that note taped up where he left it in my school mailbox.

How is it possible that the *past* doesn't exist and yet happened? It just means that events happened, but all the mass-energy of those previously organized events are with us presently today. The *past* doesn't exist in that the prior finite arrangements that made up the past are with us *Now*. And the finite arrangements we are presently experiencing *Now* are going through arrangement changes as you read this. Nothing from the *past* remains in the *past*. It is all with us in the present, *Now*. And that never changes.

Using the same logic, we can surmise that the future doesn't exist separately from the *Now*. Everything about the future is presently with us *Now*. The *future **is*** how the *present* flows. From an omnipresent state of mass-energy perspective, mass-energy is timeless, without beginning of days or end of years, so neither a *past* nor a *future* exists separately from the *present Now*.

Let's examine the perspective of *Now* using Thomas Jefferson, a historical figure of the *past*. We know that he existed. He had a body, consciousness, and left many evidences of a life well lived. Yet, where is he today? Every organized particle that made up the mass-energy of Thomas Jefferson still presently exists *Now*. If we were to go to his grave we might very well see evidence that parts of Thomas Jefferson's body remains intact? And what parts of it we didn't see doesn't mean they stopped existing. It just means that they are elsewhere in the present *Now*.

(The bodies that house our consciousness are ever-changing, finite arrangements of the infinite, never the same from one moment to the next, yet the mass-energy from which they are comprised is infinite in its duration.) From a spiritual perspective, many believe the driving life force that sparked the personality of Thomas Jefferson still exists *Now*, beyond our present ability to detect. In summary, all of the *mass-energy* that made up Thomas Jefferson throughout his entire life existed prior to his birth. And all of the *mass-energy* that made up Thomas Jefferson throughout his entire life continues to exist after his death. Thus, Thomas Jefferson was an organized arrangement of omnipresent *mass-energy*. There wasn't anything about his physical being that hasn't always existed or will not always exist as some form of *energy* confined as *mass-energies* or propagating freely throughout the universe.

Concerning the *infinite*, there is nothing *Now*, that wasn't *Then*, that won't be *Tomorrow*. That is why we treasure the *finite*.

Chapter 13

Was Einstein Wrong?

Timelessness and time dilation are contradictions!

As bought up earlier in the book, Special Relativity changed the way people think about *time*. *Time* became the fourth dimension, being weaved into the fabric of space. The basic principle of Special Relativity is not that complicated. Special Relativity is the marriage or unification of two contradicting principles, Galilean relativity and Maxwell's constant for the speed of electromagnetic waves. It is the mathematics and consequences of Special Relativity that indulges in complexity. One important consequence of Special Relativity caught our attention, *time dilation*. It stands in direct opposition to *timelessness*. If *time dilation* is a true principle, then *time* exists, and *timelessness* is void. On the other hand, if *timelessness* is an accurate explanation of the duration of mass-energy, then there is a problem with *time dilation* and Special Relativity. This was our conflict with Special Relativity and eventually, General Relativity.

When we began our quest to prove or disprove *timelessness*, we were convinced that Einstein was wrong. He gave validity to the concept of *time*, and since we were convinced that *time* does not exist, his Special and General Relativity theories must be wrong. As our theory developed and the more I understood our theory in relation to his theories, the more I realized that Einstein was on the right tract. His explanations and discoveries were brilliant. But by Einstein's own admission, he failed to resolve the contradictions between General Relativity—the laws of gravity that govern the motion of large bodies like planets—and Quantum Mechanics—the world of subatomic particles. The Missing Model of Motion bridges the gap between the two and pulls into them the physics of Isaac Newton. What I learned about Special Relativity and General Relativity is that they are accurate just as Einstein predicted. What I also learned is that they do not tell the complete story.

The main fault with Einstein's work can be explained with a simple analysis. Einstein substituted *time* and *space-time* for a quantum perspective of the momentum and movement of matter. The Missing Model of Motion replaces *time* and *space-time* and explains the movement of *mass-energy* from a quantum perspective. This corrects and continues the work of Albert Einstein.

The Fitzgerald and Lorentz Contraction Formulas

Einstein's first brilliant move was to use the Lorentz Transformation to mathematically explain Special Relativity. By using this formula, Einstein was building on factual, observed results.

Prior to Einstein's paper in 1905, Fitzgerald and Lorentz individually came up with a contraction equation to explain the results discovered by the Michelson and Morley experiment. They were mathematical formulas that provided an explanation for the lack of interference patterns as the light traveled within the Michelson and Morley apparatus. It mathematically explains the necessary contraction of the physical apparatus used by Michelson and Morley to explain the *null* result of the experiment. Einstein used this formula as the heart of his explanation for special relativity.

Here are a few quotes to help establish our position on this matter.

> Trying to account for Michelson's failure to find any movement of the earth in relation to the ether, Irish physicist George Francis Fitzgerald suggested that the measuring instruments Michelson used had contracted slightly and distorted the reading. He then produced equations showing that matter contracts in the direction of its motion, the contraction increasing as the speed increases. The Fitzgerald contraction, as this phenomenon is known is 'exceedingly small in all ordinary circumstances...If the speed is 19 miles a second – the speed of the earth around the sun—the contraction of length is 1 part in 200,000, or 2 ½ inches in the diameter of the earth. (Brian 1996, 66)

The Lorentz transformation was also a contraction formula, the formula used by Einstein for his Special Theory of Relativity.

Dutch physicist H. A. Lorentz stated that a flying charged particle foreshortened in its direction of travel would increase in mass. Einstein, in turn, applied Lorentz's equations, known as the Lorentz transformation, to all objects, including clocks and measuring instruments. Einstein showed that objects moving great speeds and over vast distances decreased in size and increased in mass. Strangest of all, he proved that at those speeds time slowed. (Brian 1996, 66))

What this boils down to is that Einstein's Special Theory of Relativity was based on a mathematical formula based on a physical reality of an actual experiment. This formula provided a basis for an explanation to the null result of the Michelson and Morley experiment. Einstein used *time* along with the Lorentz transformation equation to explain the relativity dilemma. A brilliant move since quantum mechanics as a science was in its prenatal stage. Still missing from these explanations is why *mass* moves in the first place.

By questioning the very notion of *time*, we bridged the gap between Einstein's work and quantum mechanics through the Missing Model of Motion. We asked the question that Newton should have asked: *Why do objects have momentum in the first place?* The Missing Model of Motion answers this question and replaces *time* with quantum perspective. It also explains Newton's momentum law from a quantum perspective and points out how this is the building block to quantum movement, quantum relativity, and quantum gravity through the absorption and emission of energy. Although Einstein's first two postulates (that the laws of physics are the same for all inertial frames and that the speed of light is the same for all macro moving masses), are not valid when viewed from a quantum perspective that has been introduced in this book, they were a starting point that Einstein could maneuver around with his creative use of *time*. *Time* allowed Einstein to explain the movements initiated on a quantum level from a macrophysics perspective.

Space and Time

Einstein's use of space-time to account for changes in mass-energy at different speeds provided a temporary solution to Maxwell's equations. Instead of using *time* to explain how Maxwell's constant is the same for all moving bodies, we showed how Maxwell's constant is the same for the energies that comprise any moving body. It is becomes of these energies that these bodies even move in the first place. Space-time provided a temporary substitution for quantum momentum, quantum movement, quantum relativity, and quantum gravity.

In General Relativity, Einstein's equations show a distortion of space around massive spherical bodies. In following the Missing Model of Motion to its logical consequences, I realized that gravitational energy will be greatest around the surface of massive spherical bodies and will become weaker the further this energy moves away from the surface of these massive spherical bodies. What Einstein saw as a space-time distortion was actually higher concentration of gravitational energy as explained in an earlier chapter. As with Special Relativity, Einstein used space-time as a temporary substitution for the movement of mass actually initiated on a quantum level.

Today, modern theoretical physicists are looking for clues to explain gravity by attempting to break up matter into its most elementary parts. Other physicists are creating multidimensional explanations that leave us with theories that cannot be validated. Where do we go from here? What would Einstein do? He would **continue...** to search for a quantum explanation for gravity. The Missing Model of Motion with its four legs—quantum momentum, quantum movement, quantum relativity, and quantum gravity—*continue* the work of Einstein...the rest are *details*.

Appendix 1

Gravitational Energy

○ ○

Quantum gravity is not a blind theory wherein there is no physical evidence supporting its probable validity. As an object accelerates towards the surface of the earth, it acquires energy. We see the acceleration. When it makes impact, there is a visible demonstration of the object's increased energy level as it makes smaller and smaller movements until it has reached equilibrium with the energy level of the surface upon which it finally rests. A ball bounces and bounces until it has emitted enough energy to be at balance with the surface it keeps accelerating towards. The absorption and emission of energy, which is the source and cause for all changes in momentum, are visible manifestations. We literally see the effects with our eyes. What we don't see are the exchanges taking place at the speed of light as unbalanced energy levels acquire equilibrium. Up to this point, I have used *energy* to describe the exchanges of energy that take place on a quantum level. The question I am presently unsure about is the nature of the energy absorbed and emitted by electromagnetic waves.

Are electromagnetic waves the purest form of energy in the universe?

In the early stages of trying to figure out quantum relativity, I perceived energy levels and orbital patterns as two separate processes working together. Later, I realized that changes in the frequency of electromagnetic waves described the same process. If electromagnetic waves could absorb and emit energy, their frequencies would simultaneously change. Visualizing the energy of orbital patterns or confined energies as being comprised of electromagnetic energy explained the changes in momentum through changes in frequency through the absorption and emission of energy. I realized that a change in frequency must be accompanied by a change in its energy level, causing a change in its wavelength. Frequencies do not just change without the absorption and emission of energy.

This is a very important principle. *Electromagnetic waves do not change frequencies without the absorption and emission of energy.* What energy frequency is absorbed and emitted into electromagnetic waves? The answer to this question is getting out of my philosophical realm and into a very theoretical and scientific realm somewhat beyond my present abilities to think. Electromagnetic waves are an interesting phenomenon that Maxwell brilliantly defined mathematically. Einstein defined them as quanta or packets of energy, eliminating the need for ether. The question I wonder about is how frequencies are established and how they change. This much I feel confident about: A change in frequency is a change in the actual energy confined within the electromagnetic wave. To say otherwise is to add or subtract energy from the overall energy of the universe. This means that a red shift would mean a physical loss of some energy, whereas a blue shift would mean an acquisition of energy.

We know that the smaller the frequency, the greater the quanta producing that frequency. Radio waves have less quanta or energy than gamma rays. The more energy contained in electromagnetic waves, the greater the frequency. It is a sliding scale. If you add energy, the wavelength decreases in size as the frequency increases, and visa versa, if you subtract energy, the wavelength increases in size as the frequency decreases. When an electromagnetic wave absorbs or emits energy, the result is a change in wavelength and frequency.

Now think of the momentum of *mass* as explained in this book. If electromagnetic energy confined in *mass* emitted energy, then its frequency would change to a longer wavelength while its overall *mass* would decrease. The overall effect would be a decrease in its momentum through space. If electromagnetic energy confined in *mass* absorbed energy, its wavelength would decrease in length while its overall *mass* would increase, increasing its momentum through space.

This is the physical description of *mass* as explained in Special Relativity. The mass expands or contracts, depending on the change of the wavelengths of the frequencies comprising *mass*. An increase in energy would produce shorter wavelengths, causing the *mass* to contract and visa versa, a decrease in energy would produce longer wavelengths causing the *mass* to expand. Also, an increase in energy would produce shorter wavelengths, increasing momentum, and visa versa; a decrease in energy would produce longer wavelengths, decreasing momentum.

This explains how the energy level and orbital patterns work as a single process. As electromagnetic waves confined in *mass* lose energy, the wavelengths of confined energies simultaneously increase, causing a change in the orbital patterns as explained in earlier chapters, decreasing the overall momentum of the *mass* through space. As energy is added to electromagnetic waves confined in *mass*, the wavelengths of confined energies simultaneously decrease, increasing the overall momentum of the *mass* through space. The key is electromagnetic waves confined as *mass* absorbing and emitting energies, changing in wavelength and frequency.

What kind of energy changes the frequency of electromagnetic waves through absorption and emission?

As I was playing basketball in the hot of the day, I noticed my black basketball bounced higher off the court and further away from the rim then it did the previous night as the sun was setting over the mountains. As the basketball sat in the sun, it absorbed energy. This energy affected its momentum and movement through space as explained in this book. When I picked up the black hose in my yard in the heat of the day and whipped it to create a wave like motion through the hose, the wave traveled down the length of the hose. The hose absorbed energy that affected its quantum momentum and movement. In the cool of the evening, the hose is not that pliable. Are there other energies emanating from the sun other than electromagnetic energy that can be absorbed and emitted by electromagnetic waves?

This is where my limits as a philosopher are exposed. I am not sure if the energy changing electromagnetic frequencies is within a range of electromagnetic frequencies similar to the phenomenon of the photoelectric effect or if there is a purer form of energy that can be absorbed and emitted by electromagnetic waves such as a gravitational energy. This is where mathematical proof as mentioned in chapter one will help determine.

Whatever the source of this energy, when it is absorbed it causes *mass* to accelerate. And when they are emitted from *mass*, they cause other *masses* that absorb them to accelerate towards the emission, causing what many have called the attraction of gravity. This energy is transmitted through spherical *mass* towards its surface, causing a higher concentration of this energy as is exits through the surface of spherical *mass*. This phenomenon, which Einstein called warped space-time, explains the acceleration of objects towards the surface of spherical bodies—the greater the concentration of these energies, the greater the acceleration, and visa versa. That explains why gravity is greater on Saturn and less on the moon.

What about the displacement of starlight around the sun, the validating proof for Einstein's mathematical predictions in General Relativity. Even though light travels at an extremely fast speed, when it comes close to the sun, it is exposed to the strong gravitational energy emanating from the sun for a duration of its travel because of the massive size of the sun. For this reason, the electromagnetic waves that pass near the sun absorb some gravitational *energy*, changing in frequency, causing a shift in its direction of travel. Just as electromagnetic waves confined as *mass* accelerate towards the source of energy absorption, electromagnetic waves passing near the sun would also experience the same acceleration towards the sun as it passes. Validating proof would be a shift in the frequencies of starlight observed passing near the sun. This effect would be minimized near the earth due to the earth's smaller size, decreasing the concentration of energy near its surface. The earth's smaller size also decreases the exposure time of light passing near its surface.

This book answers many questions that have perplexed science for many years. Unfortunately, it also leaves a few questions about the nature of energy such as what type of energy is absorbed into and emitted from electromagnetic waves, changing their wavelengths and frequencies. As I mentioned earlier, this reminds me of the photoelectric effect, which required the absorption of certain electromagnetic frequencies in order to create the observable effect of releasing an election from a metal plate. The effects of energy being absorbed and emitted from electromagnetic waves, whether freely moving or confined as *mass*, are clearly visible. As one mystery is unraveled, it reveals a new mystery, hidden within the unraveled mystery.

Appendix 2

Questions and Answers

During the process of developing this theory, questions would periodically pop in my mind as to how certain things were possible. After some thought concerning the matter, the answers would always come, giving further evidence that the Missing Model of Motion provides valid insight into quantum relativity, quantum momentum, quantum movement, and quantum gravity.

Question: If an object of *mass*, such as a marble, resting on the earth shares the same momentum with the earth through space, (same energy level of orbital patterns), how does it easily move in all directions on the surface of the earth?

Answer: At first, this was very difficult to visualize. I had to temporarily forget about the earth and imagine different objects moving in the same direction in space. No matter which direction you pushed them away from you, the energy level of orbital patterns would adjust to its new directional change, while still maintaining its original momentum, (unless that momentum was interfered with when its direction was changed). In other words, they would all cross the same finish line at the same time. They would just be further apart. In the same manner, because objects are accelerating towards the earth's surface as explained by quantum gravity, they are accelerating in the same direction. This allows the objects to move in any direction on the surface of the earth with the same force because they are simultaneously accelerating towards the surface of the earth or terminally velocitized to it. When *mass* appears to be resting on the surface of the earth, its energy level and orbital patterns are at terminal velocity, evenly absorbing and emitting gravitational energies. Yet, its *mass* is always juiced with the energy necessary to accelerate. Meanwhile, its energy level of orbital patterns is at equilibrium with the energy level of orbital patterns of earth's *mass*. This means that you can apply the same force in any direction to *mass* that is terminally velocitized on the surface of the earth and experience the same acceleration.

The second thing I did to help visualize this was to imagine a wall in space emitting gravitational energy. *Mass* would accelerate towards the wall. When it reached the wall, the mass would appear to rest against it. In actuality, it would be terminally velocitized against its surface. From there, the same applied force would move *mass* the same distance in any direction on the wall. In essence, the wall acts like the surface of the earth. If the *mass* was a marble and you flicked it with the same force in any direction, it would go the same distance. If you picked it up and let it go, it would accelerate towards the wall. When it hit the wall, it would bounce a few times, emitting energy until the energy level of orbital patterns of the marble matched the energy level of orbital patterns of the wall. Then it would be terminally velocitized against the surface of the wall.

Question: This one stumped me for about a day before I could visualize the answer. What if I dropped two objects at the same time, one a few inches above the other one? Wouldn't the lower object absorb most the gravitational energy, causing the second object to accelerate at a slower rate?

Answer: According to quantum gravity as introduced in this book, *mass* accelerates when gravitational energies are absorbed into it. If you could block gravitational energy, an object would appear to mysteriously float in the air. In actuality, the object's energy level of orbital patterns would be at equilibrium with that of the earth's momentum through space. It wouldn't accelerate towards earth's surface in the like manner that a razor stays next to an astronaut when he lets of it in the free falling space shuttle. In visualizing the structure of *mass*, I remembered that an atom is mostly empty space. Because of this, the part of *mass* that absorbs gravitational energies would receive very similar if not the same amount of gravitational energy without being blocked out by other atoms because of the high concentration of gravitational energies and the great amount of space occupied by an atom. It helps to visualized *mass*, such as a marble, as mostly empty space. Then visualize most of this empty space filled with gravitational energies moving through it at the speed of light. The transactions taking place transpire at the speed of light, and yet, the object accelerates at a much slower rate.

Question: If Maxwell's equations demonstrated a constant for the speed of light, how could the speed of light be constant in different inertial frames?

Answer: In understanding light moving through space, we refer back to Roemer. Roemer proved that the speed of light is finite and predictable. When Roemer made his observation of the speed of light, he accurately predicted the time it would take light to reflect off a Saturn moon and be visible at their location on earth. When earth was further away because of its orbit, the light took longer to arrive at their location in space. The duration of movement required for light to go from one of Saturn's moons to the observers on earth was a physical measurable reality. This told us that the speed of light was predictable. As we explained in an earlier chapter, the answer to the relativity dilemma did not lie in comparing the predictable speed of light to macro objects, but rather in comparing the speed of light to the energy confined as *mass*—the very source and cause of the movement of mass through space. Maxwell's constant is the same in all inertial frames from a quantum perspective. We can correctly say that electromagnetic waves travel at the same speed in all inertial frames, whether moving freely through space or confined as *mass*. This corrects Einstein's second postulate that the speed of light is the same for all moving bodies—not the macro speed of the body in whole, but the same as the energy that comprise the *mass*, the source and cause of its motion.

Question: Could gravitational energy that is absorbed and emitted by *mass* be electromagnetic waves?

Answer: See Appendix I on Gravitational Energies. Einstein was working in that direction. I think I do a good job of explaining the role of electromagnetic waves in the momentum, acceleration, and deceleration of *mass* in chapters 7 and 8.

Question: Why is there a higher concentration of energy near the surface of the earth?

Answer: Think of it this way, if atoms constantly absorb and emit energy in all directions, then a spherical shape such as the earth would emit a constant stream of energy from its surface as already explained. The earth would be constantly absorbing and emitting energy from the sun, the moon, stars, and the center of the galaxy, and even distant galaxies. The *mass* from the center of the earth outward would in essence be terminally velocitized, absorbing and emitting energy at the same rate because they are limited in movement due to constraints by opposing *masses*. Its energy level of orbital patterns would be juiced with nowhere to go so the energy would just be transmitted. The energy eventually has nowhere to go except through the surface of the earth.

Question: When a car accelerates on the surface of the earth, where does the energy come from that sustains that acceleration?

Answer: When you accelerate a car, all the orbital patterns of everything associated with the car, including the pencil on the floor, absorb the necessary energy to acquire and maintain the changing energy level of orbital patterns. As the car accelerates, all the atoms attached to it receive their increased energy from the gravitational energy being emitted from the *mass* of the earth through its surface. Remember, the only time energy is absorbed or emitted is when orbital patterns are disrupted in the moment of disruption.

Question: Concerning the displacement of stars observed close to the sun during an eclipse, does the starlight really follow a straight line through curved space, or did the gravitational energy emitted by the sun change the wavelength and frequency of starlight passing next to it, causing the starlight to accelerate towards the sun?

Answer: I believe Einstein's theory correctly predicts the displacement but that the cause will ultimately be explained from a quantum perspective. I explore this important topic in greater detail in Appendix 1 titled *Gravitational Energy*.

Question: How do less dense objects accelerate at the same rate as denser objects?

Answer: In other words, how does gravity maintain a constant acceleration for all *mass*? There is something about the make-up of contained energies in motion—*mass*—that absorbs the same ratio of energies—same acceleration—notwithstanding its atomic weight or size. Different densities of masses absorb and emit proportional amounts of energy. The proportional amount absorbed is the same proportional amount emitted at impact—the greater the mass, the greater the impact and visa versa. Nature maintains a consistent balance so the feather accelerates at the same rate as a penny in a vacuum. This is an amazing phenomenon that will yield insight into the basic structure and function of contained energies in motion. The point is even though I cannot presently explain from a quantum structure perspective how an oxygen atom accelerates at the same rate as a gold atom; the amount of energy absorbed is a fixed ratio for all contained energies in motion that is evidenced by the proportional amounts of energy released at impact, maintaining its perfect ratio.

Question: When a mirror falls to the ground, what causes it to break into many pieces?

Answer: At the moment of impact, the orbital patterns are disrupted. If the disruptions and the resulting changes in the energy level of orbital patterns are greater then the pliability of the atomic bonds that link the *mass* together, then the bonds are severed in that instant. As the mirror is falling, it is absorbing energy, and when it hits a hard surface, it is quickly emitting the same amount of energy absorbed. In the case of the mirror, this causes it to break into many fragments as equilibrium between the energy level of orbital patterns of the mirror is reestablished with that of the earth's. The momentum shifts causes bonds to sever.

Question: Why does mass accelerate in the direction of absorption?

Answer: The absorption of energy increases frequency while decreasing the wavelength of the confined energies at the point of absorption, causing a shift that accelerates the *mass* in the direction from which the energy is absorbed.

Question: Where did the idea of timelessness originate?

Answer: In 1998, Dave and I developed an understanding of omnipresence as a spiritual and philosophical continuation of ideas we were studying at the time. Omnipresence pointed towards universal *timelessness*. A discussion of the physical nature of *time* naturally resulted. This led us to *time dilation* as put forth in Einstein's Special Relativity. Our conviction that *timelessness* was not just a philosophical or spiritual idea but also a physical reality motivated our desire to question Einstein's *time dilation*. This book is the result of our collective and individual efforts. Originally, this book was divided into two parts, *Timelessness and Physics* and *Timelessness and Spirituality*. We look forward to publishing *Timelessness and the Other: Unity through Omnipresent-Living* at a later date.

Question: If the ideas presented in this book are correct, does that diminish the work of Albert Einstein?

Answer: At first, we thought it would. But as our theory developed, we realized the brilliance of Einstein's work. Our work is a continuation of his work by replacing *time* and *space-time* with a quantum perspective.

Question: How does the Missing Model of Motion unify the physics of Newton, Einstein, and Heisenberg?

Answer: The Missing Model of Motion specifically focuses on explaining how *mass* moves through space in the first place. This accounts for all movements, including Newton's laws of motion, Einstein's Relativities, and Heisenberg's work in developing quantum mechanics. None of these scientists independently explain the motion of *mass* through space in the first place. Yet, their ideas are an appendage of this phenomenon. The Missing Model of Motion connects the work of these three scientists in the momentum and movement of *mass* through space.

Question: What about Heisenberg's uncertainty principle?

Answer: I didn't delve into Heisenberg's uncertainty principle because an understanding of the Missing Model of Motion precedes the effects of the uncertainty principle. If we do not understand the role of quantum mechanics in the momentum of *mass* through space, then the uncertainty principle by itself is incomplete. Here is an interesting passage found in *Einstein, a life*:

> Because he was also corresponding with twenty-three-year-old Werner Heisenberg, Einstein was among the first to learn of his enduring contribution to quantum physics. Called "the uncertainty principle," it stated that the wave-particle duality of matter makes it impossible to determine simultaneously a particle's precise position and velocity. Consequently, investigators of the atomic and subatomic

world had entered a wonderland or crapshoot where no one could accurately predict the future from the past. (Brian, 1996, 154)

Einstein's position on quantum physics is common knowledge. It didn't sit right with him. Yet, Stephen Hawking, in referring to the unification of physics, said:

A necessary first step, therefore, is to combine general relativity with the uncertainty principle. (Hawking, 1996, 172,173)

The Missing Model of Motion is such a revolutionary idea concerning the motion of *mass* through space; it gives a whole new meaning to the function of quantum mechanics. When the role of quantum mechanics is viewed not only as being responsible for the structure of *mass*, but also for the momentum and movement of *mass* through space, then the role of the uncertainty principle will be more clear. Until then, the Missing Model of Motion precedes the uncertainty principle in unifying physics. Quantum relativity and quantum gravity are explained through an understanding of quantum momentum and quantum movement—not the uncertainty principle. Yet, somehow, I am certain that the uncertainty principle will tie into the Missing Model of Motion, providing a better understanding of its premise of the impossibility of simultaneously determining a particle's precise position and velocity.

Bibliography

Barlow, Philip L. Spring 2007. Toward a Mormon Sense of Time. *Journal of Mormon History* Vol: 33, No. 1: 1-37.

Brian, Denis. *Einstein: A Life.* New York: John Wiley & Sons, Inc. 1996.

Folger, Tim. 2000. From Here to Eternity. *Discover* December: 54-61.

Hawking, Stephen W. *A Brief History of Time.* New York: Bantam Books. 1996.

Robinson, David and Groves, Judy. *Introducing Philosophy.* London: Icon Books. 1999.

Fowler, Michael (2008). The Michelson-Morley Experiment. Retrieved August 1, 2009, Web site: http://galileo.phys.virginia.edu/classes/109N/lectures/michelson.html. (Information for Illustration—7)